Multivariate and Probabilistic Analyses of Sensory Science Problems

The *IFT Press* series reflects the mission of the Institute of Food Technologists—advancing the science and technology of food through the exchange of knowledge. Developed in partnership with Blackwell Publishing, *IFT Press* books serve as leading edge handbooks for industrial application and reference and as essential texts for academic programs. Crafted through rigorous peer review and meticulous research, *IFT Press* publications represent the latest, most significant resources available to food scientists and related agriculture professionals worldwide.

IFT Book Communications Committee

Ruth M. Patrick
Dennis R. Heldman
Theron W. Downes
Joseph H. Hotchkiss
Marianne H. Gillette
Alina S. Szczesniak
Mark Barrett
Neil H. Mermelstein
Karen Banasiak

IFT Press Editorial Advisory Board

Malcolm C. Bourne
Fergus M. Clydesdale
Dietrich Knorr
Theodore P. Labuza
Thomas J. Montville
S. Suzanne Nielsen
Martin R. Okos
Michael W. Pariza
Barbara J. Petersen
David S. Reid
Sam Saguy
Herbert Stone
Kenneth R. Swartzel

Multivariate and Probabilistic Analyses of Sensory Science Problems

Jean-François Meullenet, Rui Xiong, and Christopher J. Findlay

Jean-François Meullenet, Ph.D., is an Associate Professor in the Department of Food Science at the University of Arkansas, Fayetteville, Arkansas. Dr. Meullenet conducts research in the area of sensory science and his expertise encompasses sensory science, rheology, and modeling of food perception.

Rui Xiong, Ph.D., is a Research Scientist with Consumer Insights, Unilever Home & Personal Care, Trumbull, Connecticut.

Christopher J. Findlay, Ph.D., is President of Compusense, Inc., Guelph, Ontario, Canada. He is the Associate Editor for sensory evaluation for Food Research International.

©2007 Blackwell Publishing
All rights reserved

Blackwell Publishing Professional
2121 State Avenue, Ames, Iowa 50014, USA

Orders: 1-800-862-6657
Office: 1-515-292-0140
Fax: 1-515-292-3348
Web site: www.blackwellprofessional.com

Blackwell Publishing Ltd
9600 Garsington Road, Oxford OX4 2DQ, UK
Tel.: +44 (0)1865 776868

Blackwell Publishing Asia
550 Swanston Street, Carlton, Victoria 3053, Australia
Tel.: +61 (0)3 8359 1011

Authorization to photocopy items for internal or personal use, or the internal or personal use of specific clients, is granted by Blackwell Publishing, provided that the base fee is paid directly to the Copyright Clearance Center, 222 Rosewood Drive, Danvers, MA 01923. For those organizations that have been granted a photocopy license by CCC, a separate system of payments has been arranged. The fee codes for users of the Transactional Reporting Service is ISBN-13: 978-0-8138-0178-0/2007.

First edition, 2007

Library of Congress Cataloging-in-Publication Data
Meullenet, J.-F. (Jean-François), 1968–
    Multivariate and probabilistic analyses of sensory science problems/Jean-François Meullenet, Rui Xiong, and Christopher J. Findlay.–1st ed.
       p. cm.
    Includes bibliographical references and index.
    ISBN-13: 978-0-8138-0178-0 (alk. paper)
    ISBN-10: 0-8138-0178-8
    1. Food–Sensory evaluation–Statistical methods.   2. Multivariate analysis.   I. Xiong, Rui. II. Findlay, Christopher J.   III. Title.

    TX546.M48 2007
    664'.07–dc22

2006036135

The last digit is the print number: 9 8 7 6 5 4 3 2 1

Titles in the *IFT Press* series

- *Accelerating New Food Product Design and Development* (Jacqueline H.P. Beckley, Elizabeth J. Topp, M. Michele Foley, J.C. Huang and Witoon Prinyawiwatkul)
- *Biofilms in the Food Environment* (Hans P. Blaschek, Hua Wang, and Meredith E. Agle)
- *Food Irradiation Research and Technology* (Christopher H. Sommers and Xuetong Fan)
- *Food Risk and Crisis Communication* (Anthony O. Flood and Christine M. Bruhn)
- *Foodborne Pathogens in the Food Processing Environment: Sources, Detection and Control* (Sadhana Ravishankar and Vijay K. Juneja)
- *High Pressure Processing of Foods* (Christopher J. Doona, C. Patrick Dunne, and Florence E. Feeherry)
- *Hydrocolloids in Food Processing* (Thomas R. Laaman)
- *Microbiology and Technology of Fermented Foods* (Robert W. Hutkins)
- *Multivariate and Probabilistic Analyses of Sensory Science Problems* (Jean-François Meullenet, Rui Xiong, and Christopher Findlay)
- *Nondestructive Testing of Food Quality* (Joseph Irudayaraj and Christoph Reh)
- *Nonthermal Processing Technologies for Food* (Howard Q. Zhang, Gustavo V. Barbosa-Canovas, V.M. Balasubramaniam, Editors; C. Patrick Dunne, Daniel F. Farkas, James T.C. Yuan, Associate Editors)
- *Nutraceuticals, Glycemic Health and Diabetes* (Vijai K. Pasupuleti and James W. Anderson)
- *Packaging for Nonthermal Processing of Food* (J.H. Han)
- *Preharvest and Postharvest Food Safety: Contemporary Issues and Future Directions* (Ross C. Beier, Suresh D. Pillai, and Timothy D. Phillips, Editors; Richard L. Ziprin, Associate Editor)
- *Processing and Nutrition of Fats and Oils* (Ernesto Hernandez, Monjur Hossen, and Afaf Kamal-Eldin)
- *Regulation of Functional Foods and Nutraceuticals: A Global Perspective* (Clare M. Hasler)
- *Sensory and Consumer Research in Food Product Design and Development* (Howard R. Moskowitz, Jacqueline H. Beckley, and Anna V.A. Resurreccion)
- *Thermal Processing of Foods: Control and Automation* (K.P. Sandeep)
- *Water Activity in Foods: Fundamentals and Applications* (Gustavo V. Barbosa-Canovas, Anthony J. Fontana Jr., Shelly J. Schmidt, and Theodore P. Labuza)
- *Whey Processing, Functionality and Health Benefits* (Charles Onwulata and Peter Huth)

Table of Contents

Introduction, 3

Chapter 1. A Description of Sample Data Sets Used in Further Chapters, 9

1.1. A Description of Example Data Sets, 9
References, 25

Chapter 2. Panelist and Panel Performance: A Multivariate Experience, 27

2.1. The Multivariate Nature of Sensory Evaluation, 27
2.2. Univariate Approaches to Panelist Assessment, 29
2.3. Multivariate Techniques for Panelist Performance, 32
2.4. Panel Evaluation through Multivariate Techniques, 43
2.5. Conclusions, 46
References, 47

Chapter 3. A Nontechnical Description of Preference Mapping, 49

3.1. Introduction, 49
3.2. Internal Preference Mapping, 49
3.3. External Preference Mapping (PREFMAP), 58
3.4. Conclusions, 66
References, 67

Chapter 4. Deterministic Extensions to Preference Mapping Techniques, 69

4.1. Introduction, 69
4.2. Application and Models Available, 69
4.3. Conclusions, 89
References, 94

Chapter 5. Multidimensional Scaling and Unfolding and the Application of Probabilistic Unfolding to Model Preference Data, 95

5.1. Introduction, 95
5.2. Multidimensional Scaling (MDS) and Unfolding, 96
5.3. Probabilistic Approach to Unfolding and Identifying the Drivers of Liking, 98
5.4. Examples, 100
References, 109

Chapter 6. Consumer Segmentation Techniques, 111

6.1. Introduction, 111
6.2. Methods Available, 111
6.3. Segmentation Methods Using Hierarchical Cluster Analysis, 113
References, 126

Chapter 7. Ordinal Logistic Regression Models in Consumer Research, 129

7.1. Introduction, 129
7.2. Limitations of Ordinary Least Square Regression, 129
7.3. Odds, Odds Ratio, and Logit, 130
7.4. Binary Logistic Regression, 133
7.5. Ordinal Logistic Regression Models, 144
7.6. Proportional Odds Model (POM), 144
7.7. Conclusions, 160
References, 160

Chapter 8. Risk Assessment in Sensory and Consumer Science, 163

8.1. Introduction, 163
8.2. Concepts of Quantitative Risk Assessment, 164
8.3. A Case Study: Cheese Sticks Appetizers, 166
8.4. Conclusions, 176
References, 176

Chapter 9. Application of MARS to Preference Mapping, 179

9.1. Introduction, 179
9.2. MARS Basics, 179
9.3. Setting Control Parameters and Refining Models, 187
9.4. Example of Application of MARS, 188
9.5. A Comparison with PLS Regression, 201
References, 205

Chapter 10. Analysis of Just About Right Data, 207

10.1. Introduction, 207
10.2. Basics of Penalty Analysis, 208
10.3. Boot Strapping Penalty Analysis, 210
10.4. Use of MARS to Model JAR Data, 212
10.5. A Proportional Odds/Hazards Approach to Diagnostic Data Analysis, 215
10.6 Use of Dummy Variables to Model JAR Data, 220
References, 233

Index, 237

Multivariate and Probabilistic Analyses of Sensory Science Problems

Introduction

Multivariate analysis of sensory and consumer science data is common practice today in industry. This stems from the fact that with advances in computing and software development these analyses can be performed with ease and also because the data to be analyzed can be more complex today than in the early days of the sensory evaluation discipline. This book was written with the sensory practitioner in mind. Many in industry who are in sensory evaluation leadership roles deal with this data on a routine basis and have had, in many cases, to educate themselves about these advanced techniques while on the job. Our intent in writing this book was to provide nontechnical descriptions of multivariate techniques that are commonly and less commonly used in sensory and consumer science. The authors are not statisticians by training, which we hope helped in keeping this book at a level most sensory professionals will find comfortable. We apologize to the statisticians and sensometricians for not providing great details about statistical theories associated with these methods. However, in a few instances, we felt that some statistical details were necessary, especially when the techniques described were not widely published. We present techniques that answer specific sensory questions related to sensory and consumer testing, but we have omitted any discussions about discrimination testing as multivariate analyses are not typically used with this type of data. Table 1 summarizes the subjects this text will deal with and the corresponding sections of the book that provide answers to these questions.

Panelists and Panel Performance

Common answers provided by sensory tests dealing with descriptive analysis include quantifying sensory differences among products. This is often the case in product development when, for example, ingredients need to be substituted or the sensory profiles of competing products need to be assessed. Aspects of multivariate analysis dealing with the quantification of product differences are discussed in Chapter 2. Although descriptive analysis panels are very useful to quantify product sensory differences, they need to reach a certain level of performance to yield reliable data. Panelists and panel performance as a whole are dictated by the panelists' ability to discriminate among products when differences exist and by their ability to reproduce their assessment of the product. Both of these aspects of panel performance are important and dictate the outcome of any particular test. Several multivariate techniques have been developed to assess panelists and panel performance.

The use of MANOVA (multivariate analysis of variance) is discussed to evaluate panelists' ability to discriminate among products and panel homogeneity. To measure panel homogeneity, the use of Principal Component Analysis (PCA) and Generalized Procrustes Analysis (GPA) of individual attributes also is discussed. There are also

Table 1. Subject discussed in the text and corresponding sections.

	Chapter								
	2	3	4	5	6	7	8	9	10
Assess panelist performance	2.2 and 2.3. Univariate and multivariate panelists performance								
Assess panel performance	2.5. Examples of use of multivariate techniques								
Assess product differences	2.3. Clustering, MANOVA, PCA, GPA				6.3. and 6.4. Hierarchical clustering and latent class models by ANOVA of segments	7.6.2. Testing product differences with POM			
Understanding consumer heterogeneity		3.2. MDPREF by identification of consumer groups in a biplot	4.2.2. Response surface model by examination of density plot	5.3. Landscape Segment Analysis by examination of ideal points and density plot	6.3. Hierarchical clustering				

	PCA	GPA	MDS	Cluster analysis	POM	MARS	PLSR	Other
	3.2. Extended internal preference map for clusters	4.2.3. Euclidian distance model by examination of density plot or consumer ideals		6.4. Latent class models	7.6.3. Demographic variable effects with POM		9.4. MARS to establish relationship between OL and attribute acceptance or diagnostic attributes		10.3 Penalty and bootstrap penalty analyses
	3.3. PREFMAP by examination of score and loading plots		5.3. LSA by projection of sensory attributes in the space		7.6.5. External preference modelling with POM	8.3.1. Rejection rate from attribute acceptance			
Identifying attributes important to liking	3.2 MDPREF by projection of sensory data in the biplot					8.3.2. and 8.3.4. Rejection rate and OL from sensory attribute score distributions		9.5. PLSR to model OL from diagnostic attributes	10.4. Mars 10.5 POM and 10.6 PLSR with dummy variables to model OL from JAR data

Note: PCA = Principal Component Analysis; GPA = Generalized Procrustes Analysis; POM = Proportional Odds Model; MARS = Multivariate Adaptive Regression Splines; PLSR = Partial Least Squares Regression; OL = Overall Liking; JAR = Just About Right scales.

situations in which the objective is to compare results given by different panels evaluating the same set of products using either the same lexicon or not. Chapter 2 presents several ways to evaluate panel-to-panel consonance using the Normalized Regression Vector (NRV), PCA, or GPA.

Testing with consumers has become an integral part of a sensory scientist's weekly routine. Although this may have historically been the type of test reserved for market researchers, there has been, during the past 30 years, a slow but sure move toward consumer testing being conducted within the scope of sensory evaluation programs. We see several broad categories of questions or problems answered through testing with consumers.

Assessing Product Differences

In a typical consumer test, one is concerned with assessing differences between the products involved in the test. These differences can be related to overall liking level or to liking for a specific attribute or differences in sensory intensities measured for example on a just about right scale. Differences in liking level are commonly assessed by analysis of variance (ANOVA). Alternatively, Chapter 7 presents a probabilistic approach to comparing the distribution of hedonic scores for a series of products. This approach is based on the proportional odds model, which is preferable to ANOVA because the response variable is ordinal in nature.

Assessing the Heterogeneity of Consumer Responses

In a typical consumer test assessing liking for a series of products, consumers will not agree on the rank order of the products. This is to be expected because expectations for a product's sensory characteristics can differ quite drastically (Moskowitz and Bernstein, 2000). Failure to recognize this can yield erroneous conclusions about the test outcome. For example, two products with the same overall liking mean can show some very different distributions of liking scores. There are many approaches to evaluating consumer response heterogeneity. Early methods to assess heterogeneity of responses included MDPREF (Carroll, 1972). Internal preference mapping, as originally described, is a low multidimensional representation of liking data on a series of products. The method, described at length in Chapter 3, provides a map consisting of product locations and consumer direction of increasing liking. The degree of heterogeneity in the data is illustrated in Figures 3.2 and 3.3. In the case of Figure 3.2, the heterogeneity in consumer responses is less than that illustrated in Figure 3.3. This is because in the case of Figure 3.2, consumer vectors have a tendency to mostly point toward the right side of the figure, whereas for Figure 3.3, vectors point in all directions. Similar analysis can be performed in the external preference mapping (PREFMAP) framework, where the product locations in the map are dictated by sensory profiles and where consumers are later projected in the map.

In addition to these classic methodologies, the concept of mapping of consumer ideals in a sensory space can provide important information about consumer segmentation. The concept of ideal point has been recently implemented in probabilistic unfolding framework (Ennis et al., 1988; MacKay, 2001). In particular, we discuss in Chapter 5 a method named Landscape Segment Analysis that maps consumer ideal points and products in a common map. The examination of an ideal point density plot gives a sense for consumer segmentation. Similar deterministic methods based on response surface methodology (Danzart, 1998) and correlations between hedonic scores and distances to ideal points are discussed in Chapter 4.

A more complete identification of potential segments of consumers can be done using hedonic responses clustering, in which consumers are assigned to a group based on the pattern of liking scores given to the series of products presented. Chapter 6 of this book deals with this type of methodologies. In particular, the classical method of agglomerative hierarchical cluster analysis (Ward, 1963) is compared to latent class models (Vermunt and Magidson, 2000), which are finite mixture models and have recently been applied to the analysis of consumer data (Popper et al., 2004). Hierarchical clustering is often performed as a preprocessing step for MDPREF and PREFMAP, known as extended preference mapping which is also discussed in chapter 3.

Identifying Attributes Driving Liking

In many instances, it is not sufficient to identify what products are liked most or least, or if consumers are segmented or not. Although these considerations are important, products have to usually be optimized in a product development setting. In order to provide the product developer with clear directions for improving products, it is necessary to identify sensory attributes that are important to product liking. We discuss throughout the book several methods to do this. In chapter 3, we describe the projection of external information (i.e., sensory attributes evaluated by a trained panel) into the space. This allows the identification of attributes both positively and negatively correlated with liking. Similar interpretations are also described for external preference mapping. In chapter 4, the attributes driving liking are identified from partial least squares models determined using average data. In chapter 5, the drivers of liking are identified by projecting the sensory attributes in the LSA map. In chapter 7, we describe the application of the proportional odds model to model liking from external data while in chapter 8, we propose modeling liking score probability from a combination of partial least squares regression and distributions of sensory scores given by a trained panel.

External data use in combination with consumer test results is not always necessary or available to understand consumer test results. In some cases, consumer tests are designed so that diagnostic questions are asked in addition to hedonic questions. Diagnostic questions pertain to the appropriateness of the level or intensity of a specific attribute and are assessed using just about right scales. In chapter 9, we discuss the use of multivariate adaptive regression splines (MARS) to model the relationship between overall liking and diagnostic attributes. This technique is useful in determining if an attribute not found at a JAR level has an impact on product liking. This is useful to prioritize the attributes that may need to be optimized in a product as some attributes will not have an impact on liking even if they are not found to be at an optimal (i.e., JAR) level. In chapter 10, we discuss other methods of analysis to identify attributes driving liking. These include penalty analysis, a nonstatistical method, and bootstrap penalty analysis, which makes use of resampling techniques to determine if penalties are significantly different from zero. In addition, we further discuss the use of MARS, the implementation of the proportional odds model to model liking from JAR data, and the application of a partial least squares model with dummy variables to identify attributes driving liking. Overall, this methodology suffers from the fact that the analyst needs to compile an exhaustive list of potential drivers prior to conducting the consumer study.

Toward an Optimal Product

In many cases, it is not sufficient to determine the general role sensory attributes play toward establishing overall liking for a product. For example, knowing that sweetness positively drives liking for a product does not establish what the optimal sweetness in the product might be. This stems from the fact that more is not always better and that the relationship between liking and sensory intensity is often an inverted U-shaped curve (Moskowitz, 1985). We discuss several methods to quantify the sensory profile of an optimal product. In chapter 4, we describe a method proposed by Danzart (2004) that uses barycentric properties of existing products in the map. We propose an alternative known as the generalized inverse to determine the optimal sensory profile. Finally, we describe in chapter 5 the process of projections used in Landscape Segment Analysis (LSA) to determine the sensory profile of any virtual product in the LSA map.

The authors purposely tried to limit the number of data sets used throughout the book so that methods can be compared to each other and the reader can decide what approaches to data analysis are most appropriate for the situation at hand. We hope that by providing access to these data sets through the web site that accompanies this book, some of the readers might feel more comfortable in reproducing the analyses performed as illustrations throughout this book. In addition, we certainly hope that some of you will make improvements to the methods presented here and provide significant advancements to the field of sensory and consumer sciences.

References

Carroll, J.D. 1972. Individual differences and multidimensional scaling. In R.N. Shepard, A.K. Romney, and S.B. Nerlove (Eds.) Multidimensional Scaling: Theory and Applications in the Behavioral Sciences, Vol. 1 (pp. 105–155). New York: Seminar Press.
Danzart, M. 1998. Quadratic model in preference mapping. 4th sensometric meeting, Copenhagen, August 1998.
Danzart, M., Sieffermann, J-M, and Delarue, J. 2004. New developments in preference mapping techniques: Finding out a consumer optimal product, its sensory profile and the key sensory attributes. 6th Sensometric Meeting, Davis, CA, USA, August 2004.
Ennis, D.M., Palen, J., and Mullen, K. 1988. A multidimensional stochastic theory of similarity. Journal of Mathematical Psychology, 37, 104–111.
MacKay, D.B. 2001. Probabilistic unfolding models for sensory data. Food Quality and Preference 12:427–436.
Moskowitz, H.R. 1985. New Directions in Product Testing and Sensory Analysis of Food. Westport: Food and Nutrition Press.
Moskowitz, H.R. and Bernstein, R. 2000. Variability in hedonics: Indications of worldwide sensory and cognitive preference segmentation. J. Sensory Studies, 15, 263–284.
Popper, R., Kroll, J., and Magidson, J. 2004. Application of latent class models to food product development: a case study. Sawtooth Software Conference Proceedings, 2004.
Vermunt, Jeroen K. and Magidson, J. 2000. Latent class cluster analysis. In J.A. Hagenaars and A.L. McCutcheon, (Eds.), Advances in Latent Class Models. Cambridge, UK: Cambridge University Press.

1 A Description of Sample Data Sets Used in Further Chapters

1.1. A Description of Example Data Sets

The various data sets described here are used in various chapters of the book, but we thought it would be useful to describe the methodologies in a little more detail here. The actual data can be downloaded from the book Web site at http://www.uark.edu/ua/multivariate/index.htm.

1.1.1. White Corn Tortilla Chips

1.1.1.1. Introduction

This data set is used extensively as an example in Chapters 3, 4, 5, and 7. Additional details about these studies are published in Meullenet et al. 2002 and 2003. Households in the United States are very fond of salty snacks, including tortilla chips. According to Dies (2000), 76% of U.S. households purchase tortilla chips every 32 days. In 1996, tortilla and potato chips accounted for five of the top 10 products in the salty snack category (Anonymous, 1998). In the United States, tortilla chips have a 20% market share of salty snack purchases, second only to potato chips (Lisser, 1993). In 1996, tortilla chips enjoyed their highest sales in 10 years and yellow tortilla chips gained in popularity (Wellman, 1997).

A few studies have been carried out for the evaluation of the sensory properties of tortilla chips. Buttery and Ling (1995, 1998) characterized the flavor volatiles of corn tortilla chips, whereas Hawrysh et al. (1995) examined the sensory and chemical stability of tortilla chips fried in canola oil, corn oil, and partially hydrogenated soybean oil. Stinson and Tomassetti (1995) showed that the flavor and texture acceptability of low-fat tortilla chips increased when natural corn flavor was added. However, the consumer preference pattern of white corn tortilla chips is not yet clearly defined. This study was conducted to determine the preferences for tortilla chips and to quantify the specific sensory characteristics found in the tortilla chips presently on the market.

1.1.1.2. Samples

A broad range of 25 commercially available tortilla chips were purchased at local supermarkets. The original group of tortilla chips, which included both yellow and white corn chips, was visually screened based on color, shape, and levels of salt and fat reported on the labels. Eleven commercially available toasted white corn tortilla chip products (Table 1.1) were selected for this study because white corn tortilla chips are more popular with consumers than yellow corn tortilla chips and there was a financial constraint for this project. Large bags of each of the 11 types of tortilla chips were purchased 1–2 days before testing. The samples were randomly coded with a three-digit number and stored to prevent fractioning of chips.

Table 1.1. Commercial pure white corn tortilla chip products.

Product Name	Abbreviation	Producer	Shape	Salt Content (%)[1]	Fat Content (%)[1]
Best Yet White Corn	BYW	Fleming Companies, Inc.	Triangle	4	12
Green Mountain Gringo	GMG	Green Mountain Gringo	Strip	5	13
Guy's Restaurant Rounds	GUY	Guy's Snack Foods	Round	3	9
Medallion White Corn	MED	Medallion Food Corporation	Triangle	2	11
Mission Strips	MIS	Mission Food Corporation	Strip	4	10
Mission Triangle	MIT	Mission Food Corporation	Triangle	4	10
Oak Creek Farms—White Corn	OAK	Oak Creek Farms	Round	2	11
Santita's	SAN	Frito-Lay	Triangle	5	8
Tom's White Corn	TOM	Tom's Foods Inc.	Triangle	5	10
Tostito's Bite Size	TOB	Frito-Lay	Round	5	12
Tostito's Restaurant Style	TOR	Frito-Lay	Triangle	3	9

[1]Expressed as percentage daily intake; percentage daily value is based on a 2000-calorie diet.

The coded samples were presented to panelists on white plastic plates that were 6 inches in diameter. Approximately five or six large chips or six to eight bite-size chips were placed on each plate. Each bag of chips was immediately resealed using a bag clip to preserve freshness. Chips that were excessively fractured were discarded.

1.1.1.3. Descriptive Analysis

Eleven tortilla chip samples were evaluated for appearance, flavor, and texture by a group of nine Spectrum-trained panelists. Panel orientation was conducted to develop a descriptive lexicon for the appearance, flavor, and texture attributes specific to tortilla chips, using the Spectrum method (Sensory Spectrum Inc., Chatham, NJ). References used for panel orientation and lexicon development are presented in Tables 1.2, 1.3, and 1.4. The panel orientation on tortilla chip texture, flavor, and appearance was conducted over three 3-hour sessions. The texture ballot analyzed four major categories of product texture characteristics—surface, first bite, chew down, and residual—for a total of 15 texture attributes (Table 1.2). The flavor, basic taste, aromatics, feeling factors, and aftertaste profiles consisted of 23 attributes (Table 1.3). The appearance ballot consisted of five attributes (Table 1.4).

Texture, flavor, and appearance evaluations were carried out under controlled conditions over two 3-hour sessions, two 2.5-hour sessions, and one 2-hour session, respectively. During each session, the samples were presented in a random order to each panelist (i.e., each panelists was assigned a different randomization scheme). The evaluation was performed in individual booths featuring controlled lighting and positive air pressure. Panelists were provided with a paper ballot, references, and crackers (Nabisco Premium Unsalted brand) and water for cleansing and rinsing their palates between each sample. A 10-minute break was scheduled during each session. The texture and flavor evaluations

Table 1.2. Texture vocabulary for tortilla chips.

Term[1]	Definition	Technique	Reference
Surface Characteristics			
Roughness	The amount of irregular particles in the surface of the sample: Micro: the small crevices and irregularities on the surface of the sample. Macro: The large peaks and sloped angles detected on the surface of the sample.	Hold sample to mouth and feel the surface of the sample with tongue and lips.	Graham cracker Micro 1.0 Macro 1.5 Club cracker Micro 3.5 Macro 2.5 Cookie Micro 0.8 Macro 5.0 Pringles Micro 8.0 Macro 1.0 Twizzler Micro 0.5 Macro 7.0 Granola bar Micro 12.0 Macro 1.0 Triscuit Micro 13.0 Macro 4.0 Rye wafer Micro 15.0 Macro 0.0 Nutrageous Micro 14.0
Oily/greasy lips	The amount of oily or greasy residue (regardless of the thickness) felt on the lips after placing the sample in mouth.	Place sample between lips, compress, and release. Using the tongue to feel the surface of the lips, evaluate the amount of oily/greasy residue felt on the lips.	Club cracker 5.0 Pringles 8.0
Loose particles	The amount of particles felt on the lips.	Compress sample lightly between lips, remove sample and evaluate.	Pringles 5.0
First Bite			
Hardness	The force required to compress the sample.	Compress or bite through sample one time with molars or incisors.	Cream cheese 1.0 Egg white 2.5 American cheese 4.5 Beef frankfurter 5.5 Olive 7.0 Peanut 9.5 Almond 11.0 Carrots 11.0 LifeSavers 14.5
Crispiness	The amount of small breaks felt (perceived as having many light, airy, small breaks), and the	Compress sample with molar teeth until sample shatters, breaks or crumbles.	Granola bars 2.0 Club cracker 5.0 Graham cracker 6.5 Cheerios 7.0

(*Continued*)

Table 1.2. *Continued.*

Term[1]	Definition	Technique	Reference
	degree of pitch and sound heard when the sample is cracked, broken, or compressed once.		Bran flakes 9.5 Goldfish crackers 11.0 Cornflakes 14.0
Chew Down Characteristics			
Cohesiveness of mass	The amount that the chewed sample holds together.	Chew sample with molar teeth 8–10 times and evaluate.	Shoestring licorice 0.0 Carrots 2.0 Mushrooms 4.0 Beef frankfurter 7.5 American cheese 9.0 Brownie 13.0 Dough 15.0
Moistness of mass	The amount of wetness/oiliness felt on the surface of the mass.	Chew sample with molar teeth 8–10 chews and evaluate. How wet does the sample feel?	Saltine crackers 4.5 Pound cake 10.5 Jell-O gelatin 14.0 Water 15.0
Roughness of mass	The amount of roughness perceived in the chewed sample.	Chew sample with molars 8–10 times and evaluate the irregularities in the sample mass.	Unchewed Jell-O gelatin 0.0 Orange peel 3.0 Cooked oats 6.5 Pringles 8.0
Moisture absorption	The amount of saliva absorbed by the sample during mastication. Pay attention to how dry your mouth is becoming, not to the sample.	Chew the sample with molar teeth 8–10 times and evaluate.	Shoestring licorice 0.0 Licorice 3.5 Popcorn 7.5 Potato chips 10.0 Pound cake 13.0 Saltine crackers 15.0
Persistence of crisp	The amount of mastication before the "crisp sound" changes.	Chew sample with molars and count the number of chews completed before the pitch or crisp sound changes.	
Residual Characteristics			
Oily/greasy film	The amount and degree of residue felt by the tongue when moved over the surface of the mouth.	Expectorate the sample and feel the surface of the mouth with tongue to evaluate.	Saltine crackers 0.5 Ritz crackers 2.5 Pringles 5.0 Oil 15.0
Toothpack	The amount of product packed into the crowns of your teeth after mastication.	Chew sample 15–20 times, expectorate and feel the surface of the crowns of the teeth to evaluate.	Captain Crunch cereal 5.0 Heath Bar 10.0
Loose particles	The amount of particles remaining in and on the surface of the mouth after swallowing.	Chew sample with molars, swallow and evaluate.	Carrot 10.0

[1]The data for roughness, hardness, cohesiveness of mass, roughness of mass, toothpack, and loose particles were from the work by Meullenet and others (1999, 2001); those for moisture of absorption was from the work by Meilgaard et al. (1999); the remaining data were from this study.

Table 1.3. Flavor vocabulary for tortilla chips.

Term	Definition	Reference	Intensity
Basic Tastes[1]			
Sweet	The basic taste, perceived on the tongue, stimulated by sugars and high-potency sweeteners.	Solutions of sucrose in spring water.	Sucrose solution 2%: 2.0 Sucrose solution 5%: 5.0 Sucrose solution 10%: 10.0 Sucrose solution 16%: 15.0
Salt	The basic taste, perceived on the tongue, stimulated by sodium salt, especially sodium chloride.	Solutions of sodium chloride in spring water.	NaCl solution 0.2%: 2.0 NaCl solution 0.35%: 5.0 NaCl solution 0.5%: 8.5 NaCl solution 0.55%: 10.0 NaCl solution 0.7%: 15.0
Sour	The basic taste, perceived on the tongue, stimulated by acids, such as citric acid.	Solutions of citric acid in spring water.	Citric acid solution 0.05%: 2.0 Citric acid solution 0.08%: 5.0 Citric acid solution 0.15%: 10.0 Citric acid solution 0.20%: 20.0
Bitter	The basic taste, perceived on the tongue, stimulated by substances such as quinine, caffeine, and certain other alkaloids.	Solutions of caffeine in spring water.	Caffeine solution 0.05%: 2.0 Caffeine solution 0.08%: 8.0 Caffeine solution 0.15%: 10.0 Caffeine solution 0.20%: 15.0
Aromatics[2]			
Grain complex	The overall grain impact.		Saltine: 3.0 Applesauce: 7.0 Orange juice: 10.0 Grape juice: 14.0 Big Red brand gum: 16.0
Raw corn	The aromatic associated with fresh or uncooked corn.	Cornmeal paste	
Toasted corn	The aromatic associated with caramelized or browned corn meal.	Corn Chex cereal	
Masa	The aromatic associated with alkali-treated cornmeal.	Fresh white corn tortillas; Tostitos-brand tortilla chips	
Toasted grain	A general term used to describe the aromatics of toasted grains, which can not be tied to a specific grain source.	Multiple-grain breakfast cereals	
Heated oil	The aromatic associated with fresh oil that is heated.	Fresh, heated Crisco-brand vegetable oil	
Scorched	The aromatic associated with overheating or overcooking.	Scorched popcorn	
Cardboard (papery/packaging)	Aromatics associated with slightly oxidized fats and oils or by other sources, such as staling flours; reminiscent of wet cardboard packaging.	Wet cardboard presented in a reference jar	

(*Continued*)

Table 1.3. *Continued.*

Term	Definition	Reference	Intensity
Feeling Factors[3]			
Astringent	A chemical feeling factor felt on the tongue or other surfaces of the oral cavity, described as puckering/dry and associated with tannins or alums.	Plain yogurt	Plain yogurt: 3.5

[1]The reference and intensity for basic taste was the one used by Meilgaard et al. (1999).
[2]The reference and intensity for aromatics was a slightly modified version of the reference and intensity used by Meilgaard et al. (1999).
[3]The reference and intensity for feeling factor was the one used by Meullenet et al. (2001).

Table 1.4. Appearance vocabulary for tortilla chips.

Term[1]	Definition	Technique	Reference
Degree of whiteness	The amount of pure whiteness in the sample	Observe all the chips on the tray	Nacho Rounds: 0 Green Mountain Gringo: 2 Guy's Restaurant Style: 7 White plate: 15
Grain flecks	The amount (regardless of size) of dark colored particles that appear to be coarsely ground grain particles	Observe all the chips on the tray	Green Mountain Gringo: 3 Tom's White Corn: 6 Best Yet White Corn: 8
Char marks	The amount of charred markings on the surface of the sample (usually on one side) that appears to be charred from the baking surface	Observe all the chips on the tray	Tostito's Restaurant Style: 0 Mission Strips: 5 Green Mountain Gringo: 7
Micro surface particles	The amount of fine particles laying on the surface of the sample that appear to be grain dust or salt	Observe all the chips on the tray. Turn chip sideways and examine	Green Mountain Gringo: 2 Tostito's Bite Size: 5 Oak Creek Farms: 9
Amount of bubbles	The amount of surface area of the sample affected by bubbles	Observe all the chips on the tray. Estimate the proportion of the surface area of the sample affected by bubbles	None: 0 Entire surface affected: 15

[1]Degree of whiteness was from the work by Meullenet et al. (2001); the remaining scales were developed for this study.

were performed in duplicate during two sessions held on different days. The evaluation of product appearance was performed by presenting 10 chips during a single session. The panelists were asked to evaluate the appearance of all the chips and to give an average score for each attribute. This method was used because of the large variation in appearance within each product.

1.1.1.4. Consumer Testing

Based on the work of Dies (2000), who reported that 76% of U.S. households purchase tortilla chips every 32 days, we assumed that 75% of the population in Northwest Arkansas consumes tortilla chip products on a regular basis. The minimum sample size necessary to obtain a representative sample of the population was calculated to be 73 consumers, using the formula $Z_\alpha p(1-p)/C_p^2$ (Rea and Parker, 1992; in this case, $Z_\alpha = 1.96$, associated with 95% confidence level, the proportion $p = 0.75$, and the confidence interval $C_p^2 \leq 10\%$). Because the consumer test was performed over a 2-day period, 80 consumers were selected in anticipation of second-day no shows.

The consumer panel was recruited by posting advertisements at local restaurants, grocery stores, supermarkets, and the university campus, requesting participation in a salty snack consumer taste. Consumers of these snack products between the ages of 18 and 35 years who were interested in participating were asked to call a number given on the advertisements. A phone screener was used to determine whether the caller consumed tortilla chips on a regular basis, which was considered to be every 2 weeks, or twice a month. Eighty tortilla chip consumers were selected and scheduled for the 2-day test. A $20 gift card was offered as an incentive and was paid on completion of the test.

A completely randomized design was used across the 11 samples for the 80 consumers (Meilgaard et al., 1999). The 2-day consumer test was carried out at the University of Arkansas Sensory Laboratories. Each consumer was seated in an individual testing booth with controlled lighting and positive airflow and was presented with five and six tortilla chip samples on the first and second days of the test, respectively. The consumers were provided with five or six large chips or from six to eight bite-size chips for each sample, presented on 6-inch-diameter coded white plastic plates. Each sample was assigned a three-digit code to be entered by consumers on the ballot as a means of identifying the sample. Each consumer was asked to evaluate the appearance, overall impression, flavor, and texture of each sample on a 9-point hedonic scale with 9 being "like extremely" and 1 being "dislike extremely" (Table 1.5). Consumers were also asked to rate the amount of saltiness on a 5-point Just About Right scale with Just About Right being a score of 3 (1 = "much too low" and 5 = "much too high"; Table 1.6). Demographic data including gender, age group, consumption frequency, and preference for brands and shapes of tortilla chips were also gathered.

1.1.1.5. Appearance Measurements

The color of the samples was instrumentally evaluated using a Minolta CR-300 colorimeter (Minolta Co., Ltd., Osaka, Japan). The hunter color values (L, a, and b) were used. The color value "L," measuring whiteness, is quantified on a scale from 0 to 100. The color value "a" quantifies red (positive values) to green (negative values), and the color value "b" quantifies yellow (positive values) to blue (negative values).

1.1.1.6. Summary Results

The frequency (percentage) of the consumer overall acceptance and acceptance of appearance, flavor, texture, and amount of salt is presented in Table 1.6. It was evident that the Tostito's Restaurant Style chips had the highest overall acceptance frequency of "Like extremely" responses, followed by the Tostito's Bite Size chips. Both Tostito's Bite Size and Tostito's Restaurant Style tortilla chips, together with Santita's chips, from the same manufacturer (Frito-Lay), appeared to have the highest percentages (57.6%, 47.5%, and

Table 1.5. Means of descriptive and /or instrumental attributes for appearance, flavor and texture

Attribute	White Corn Tortilla Chips[1]									
	BYW	GMG	GUY	MED	MIS	MIT	OAK	SAN	TOB	TOM	TOR
Appearance										
Instrumental color L	95.9	92.7	93.7	84.9	68.9	69.6	96.5	93.1	95.9	92.9	92.7
Instrumental color a	−0.1	−0.3	−0.6	0.4	0.4	0.3	0.6	−0.8	−0.4	−0.3	−0.8
Instrumental color b	12.1	22.1	2.6	9.5	23.5	23.6	7.4	8.8	13	8.7	4.9
Degree of whiteness	6	2	7	6.5	5	3	6	6	6	7	6
Grain flecks	8	3	6	3.5	4	3	6	8	6	6	8
Char marks	2	7	7	3	5	4.5	2	1.5	0	6	0
Micro surface particles	2	0.5	1	2	2.5	2	2	1	2.5	3.5	2
Amount of bubbles	6	6	6	6	5	5	6	7	7.5	7	5
Flavor										
Sweet	0.5	0.5	0.5	0.4	0.6	0.4	0.7	0.6	0.4	0.4	0.6
Salt	8.8	7.2	7.8	6.9	9.4	9.9	7.3	9	8.9	8.5	8
Grain complex	6.9	7.2	6.6	6.5	7.1	6.9	6.8	7	6.9	7	6.9
Toasted corn	2.7	2.3	1.6	2.3	3.8	2.8	1.3	3.4	2.6	2.5	2.6
Raw corn	0	0.2	0	0	0	0	0.3	0	0	0	0
Masa	3.7	3.7	3.8	3.8	3.1	3.2	4.2	3.7	4	4.1	3.4
Toasted grain	1.4	2.2	2.2	1.5	1.2	2	1.8	0.8	1	1.4	2.3
Sweet	0.5	0.5	0.5	0.4	0.6	0.4	0.7	0.6	0.4	0.4	0.6
Heated oil	4.4	4.4	4.5	4.3	4.5	4.4	4.2	4.3	4.5	4.6	4.3
Scorched	0	0.5	0.6	0	2.7	0.6	0	0	0.2	0	0
Cardboard	2.4	2	2.8	3.3	2.7	2.6	2.6	2.1	1.9	2.4	2.4
Astringent	2.6	2.5	2.5	2.5	2.7	2.6	2.6	2.6	2.6	2.6	2.5
Aftertaste: toasted corn	1.4	1.1	0.7	1.1	2.2	1.8	0.5	2.4	1.7	1.1	1.5
Aftertaste: toasted grain	1	1.4	1.2	1.2	0.8	0.9	1.2	0.4	0.4	1.1	1.4
Aftertaste: toasted masa	2	1.6	2	1.5	1.3	1.7	1.5	1.6	1.3	1.7	1.2
Aftertaste: raw masa	0.2	0.9	0.6	0.7	0.2	0	1.3	0.2	0.4	1.1	0.6
Aftertaste: heated oil	2.5	2.8	2.8	2.7	3.1	2.7	2.4	2.8	3.1	2.8	2.9
Aftertaste: salt	3.6	3.1	3.3	2.9	3.9	4.1	3.3	3.8	4.1	3.5	3.7
Aftertaste: sweet	0.4	0.4	0.5	0.3	0.4	0.4	0.5	0.4	0.3	0.3	0.4
Texture										
Micro roughness	8.8	7.6	8.6	8.3	8.1	8.7	8.2	7.6	7.8	6.9	7.7
Macro roughness	3.2	3.1	2.8	4.4	4.1	4.3	3.2	4.1	3.8	5.5	4.4
Oily/greasy lips	6.7	4.4	5	6	5.3	5.8	6	5.8	5.8	5.9	5.3
Loose particles	6.7	4.4	5	6	5.8	5.3	6	5.8	5.8	5.9	5.3
Hardness	8.8	9.6	8.6	9	8.7	8.8	8.7	8.8	8.2	8.5	8
Crispness	10.4	9.6	9.9	10.6	10.2	10.2	9.5	10.7	11.1	10.7	11.5
Fracturability	7.4	7.2	7.8	7.9	7.5	7.7	7.1	7.7	8	7.5	7.9
Cohesiveness of mass	2.7	3.3	3.2	3.1	2.9	3	3.5	3	3.2	3.1	3.4
Roughness of mass	7.6	7.2	7.4	7.5	7.3	7.5	7.5	7.6	7.3	7.7	7.6
Moistness of mass	7.3	7.6	7.1	7.2	7.3	6.9	7.7	7.4	7.3	7.7	7.2
Moisture absorption	9.1	9.2	9.5	9.1	9.6	9.7	9.1	9.5	9.1	9	9.3
Persistence of crisp	6	5.2	5.1	5.5	4.9	5.3	5.1	5.7	5.7	5.4	5.4
Toothpack	5.2	5.2	5.1	5.2	5.4	5.8	5.4	5.3	5.3	5.2	5.1
Loose particles	7.2	7.2	6.9	7	7.3	7	7.1	7.5	7.3	7.3	6.9
Oily/greasy film	4	3.8	3.9	4.3	3.9	3.9	3.8	3.8	3.9	4.5	3.8

[1]Sample name abbreviations can be found in Table 1.1.

Table 1.6. Frequency (%) of consumer overall acceptance and consumer acceptance of appearance, flavor, texture and saltiness for 11 tortilla chip products[1].

Scale[2]	White Corn Tortilla Chips[3]										
	BYW	GMG	GUY	MED	MIS	MIT	OAK	SAN	TOB	TOM	TOR
Overall acceptance											
Dislike extremely (1)	0.0	5.0	1.3	3.8	1.3	0.0	6.3	0.0	0.0	5.0	0.0
Dislike very much (2)	0.0	5.0	2.5	7.5	2.5	3.8	3.8	0.0	0.0	1.3	2.5
Dislike moderately (3)	3.8	8.8	6.3	7.5	6.3	2.5	6.3	0.0	3.8	3.8	1.3
Dislike slightly (4)	11.3	13.8	17.5	18.8	12.5	10.0	17.7	7.5	3.8	6.3	2.5
Neither dislike nor like (5)	13.8	10.0	7.5	8.8	3.8	5.0	7.6	6.3	3.8	12.5	5.0
Like slightly (6)	13.8	23.8	21.3	20.0	18.8	8.8	16.5	7.5	12.5	12.5	15.0
Like moderately (7)	30.0	16.3	22.5	25.0	26.3	36.3	27.8	33.8	18.8	30.0	26.3
Like very much (8)	26.3	13.8	20.0	6.3	22.5	25.0	10.1	38.8	43.8	21.3	30.0
Like extremely (9)	1.3	3.8	1.3	2.5	6.3	8.8	3.8	6.3	13.8	7.5	17.5
Appearance											
Dislike extremely (1)	0.0	1.0	0.0	0.0	0.0	0.0	1.0	0.0	0.0	1.0	0.0
Dislike very much (2)	0.0	0.0	1.0	4.0	3.0	0.0	2.0	0.0	1.0	1.0	2.0
Dislike moderately (3)	3.0	4.0	5.0	2.0	6.0	4.0	1.0	0.0	2.0	0.0	1.0
Dislike slightly (4)	16.0	11.0	10.0	13.0	15.0	7.0	6.0	2.0	6.0	6.0	7.0
Neither dislike nor like (5)	4.0	11.0	6.0	8.0	6.0	5.0	11.0	5.0	4.0	6.0	3.0
Like slightly (6)	13.0	12.0	14.0	9.0	13.0	12.0	10.0	8.0	6.0	13.0	9.0
Like moderately (7)	21.0	19.0	22.0	27.0	12.0	20.0	27.0	26.0	16.0	19.0	19.0
Like very much (8)	21.0	17.0	18.0	13.0	15.0	24.0	12.0	32.0	29.0	27.0	27.0
Like extremely (9)	2.0	4.0	4.0	4.0	10.0	8.0	9.0	7.0	16.0	7.0	12.0
Flavor											
Dislike extremely (1)	0.0	5.1	1.3	7.5	2.5	0.0	8.9	0.0	0.0	1.3	0.0
Dislike very much (2)	0.0	7.6	1.3	10.0	3.8	1.3	11.4	0.0	0.0	5.0	2.5
Dislike moderately (3)	6.3	7.6	8.8	8.8	7.5	6.3	6.3	1.3	2.5	5.0	0.0
Dislike slightly (4)	13.8	16.5	22.5	18.8	6.3	6.3	19.0	6.3	3.8	16.3	6.3
Neither dislike nor like (5)	10.0	12.7	8.8	6.3	6.3	8.8	10.1	7.5	3.8	15.0	3.8
Like slightly (6)	13.8	17.7	21.3	22.5	23.8	15.0	13.9	8.8	17.5	12.5	18.8
Like moderately (7)	22.5	16.5	17.5	15.0	21.3	25.0	17.7	27.5	21.3	16.3	20.0
Like very much (8)	28.8	12.7	16.3	10.0	21.3	27.5	8.9	36.3	33.8	21.3	35.0
Like extremely (9)	5.0	3.8	2.5	1.3	7.5	10.0	3.8	12.5	17.5	7.5	13.8
Texture											
Dislike extremely (1)	0.0	3.8	0.0	1.3	0.0	0.0	1.3	0.0	0.0	0.0	0.0

(Continued)

Table 1.6. *Continued.*

Scale[2]	White Corn Tortilla Chips[3]										
	BYW	GMG	GUY	MED	MIS	MIT	OAK	SAN	TOB	TOM	TOR
Dislike very much (2)	1.3	6.3	0.0	7.5	1.3	1.3	2.6	0.0	0.0	3.8	1.3
Dislike moderately (3)	2.5	6.3	8.8	1.3	3.8	5.0	5.1	0.0	0.0	2.5	1.3
Dislike slightly (4)	12.5	26.6	16.3	18.8	5.0	15.0	7.7	3.8	2.5	11.4	0.0
Neither dislike nor like (5)	1.3	8.9	11.3	8.8	8.8	5.0	9.0	7.5	5.1	15.2	6.3
Like slightly (6)	12.5	15.2	13.8	15.0	17.5	15.0	19.2	13.8	16.5	11.4	8.8
Like moderately (7)	28.8	21.5	27.5	27.5	22.5	25.0	28.2	23.8	17.7	21.5	23.8
Like very much (8)	31.3	10.1	20.0	17.5	31.3	25.0	21.8	35.0	38.0	27.8	38.8
Like extremely (9)	10.0	1.3	2.5	2.5	10.0	8.8	5.1	16.3	20.3	6.3	20.0
Amount of Salt											
Much too low (1)	3.8	20.0	8.8	31.3	1.3	5.0	10.1	0.0	2.5	8.8	3.8
Low (2)	31.3	46.3	41.3	41.3	10.1	10.0	44.3	12.5	22.5	32.5	23.8
Just right (3)	48.8	32.5	36.3	15.0	63.3	60.0	38.0	62.5	62.5	46.3	65.0
High (4)	13.8	1.3	13.8	12.5	20.3	18.8	6.3	22.5	12.5	11.3	7.5
Much too high (5)	2.5	0.0	0.0	0.0	5.1	6.3	1.3	2.5	0.0	1.3	0.0

[1]Total observations (*n* = 80 consumers) for each tortilla chip sample.
[2]A 9-point hedonic scale for consumer overall acceptance and acceptance of appearance, flavor, and texture; a 5-point Just About Right scale for the amount of salt (saltiness).
[3]Sample name abbreviations can be found in Table 1.1.

45.1%, respectively) of "Like" to "Like extremely," whereas the Medallion White Corn, Green Mountain Gringo, and Oak Creek Farms—White Corn chips had the lowest percentages of these responses (8.8%, 17.6%, and 13.9%, respectively). The acceptance of the appearance, flavor, and texture of Tostito's Bite Size chips was rated as "Like extremely" by 16.0%, 17.5%, and 20.3% of the consumers, respectively. In contrast, the acceptance of the appearance, flavor, and texture of the Medallion White Corn chips was rated as "Like extremely" only by 4.0%, 1.4%, and 2.5% of the consumers, respectively. In terms of the amount of salt, 60%–65% of the consumers rated five products (Mission Strips, Mission Triangle, Santita's, Tostito's Bite Size, and Tostito's Restaurant Style chips) as Just About Right, whereas Medallion White Corn chips had the lowest rate of approval (i.e., 15%). It was found that the Just About Right rates were associated with the salt content (Table 1.6). A salt content of 4% was found to be Just About Right for the consumer saltiness acceptance.

Descriptive intensity means for the attributes of visual appearance, flavor, and texture and the means of instrumental color values ("L," "a," and "b") are presented in Table 1.5, illustrating some of the sensory differences observed among the 11 products.

1.1.2. Muscadine Grape Juices

1.1.2.1. Introduction

These data are used in Chapter 3 on preference mapping and in Chapter 6 on consumer segmentation. Muscadine grapes (*Vitis rotundifolia*) are native to the southeastern United

States and have been produced commercially in Arkansas since 1972 (Lanier and Morris, 1979). Recent economic analysis indicates that muscadine grape production can be a profitable enterprise for growers (Noguera et al., 2005). Interest in production of muscadines has increased since muscadines and products from muscadines were identified as a source of vitamins, minerals, fiber, antioxidants, and phenolic compounds (Ector, 2001; Ector et al., 1996; Lee and Talcott, 2004; Pastrana-Bonilla et al., 2003; Striegler et al., 2005; Talcott and Lee, 2002; Threlfall et al., 2005; Yilmaz and Toledo, 2004).

The increase in consumer interest in a healthier lifestyle provides an opportunity for the muscadine industry to capture a section of the nutraceutical market that is projected to reach $28 billion in the United States by 2006 (Green, 2003). The muscadine grape juice used to study the acceptability and sensory properties of the product was hand-harvested from black and bronze cultivars from the University of Arkansas Southwest Research and Extension Center. The cultivars included Black Beauty (Brooks and Olmo, 1991), Carlos (Brooks and Olmo, 1970), Granny Val (Brooks and Olmo, 1991), Nesbitt (Goldy and Nesbitt, 1985), Southern Home (Mortensen et al., 1994), Summit (Lane, 1977), and Supreme (Brooks and Olmo, 1991).

1.1.2.2. Samples

Thirty-six kilograms of each grape cultivar were used for juice processing. The grapes were crushed twice and placed in a plastic container with a food-grade polyethylene liner. After being processed, the juice was placed in cold storage overnight and then stabilized for 2 weeks at 2°C. The juice was then pasteurized and sealed in glass bottles and stored again at the same temperature.

Each muscadine juice was assigned a random three-digit number for identification. Two commercial muscadine juices were also included, one red juice and one white juice.

1.1.2.3. Descriptive Analysis

The 10 muscadine juices were examined for aroma, basic tastes, aromatics, feeling factors, and aftertaste by a group of nine Spectrum-trained panelists. Panel orientation was held over a 2-day period during two 3-hour sessions to develop a descriptive lexicon. The lexicon and references used to describe the sensory properties of muscadine grape juice can be found in Table 1.7.

Flavor, aroma, feeling factors, and aftertaste were evaluated under controlled conditions over two sessions of 3 hours each. The samples were replicated and presented to each panelist according to different randomized schemes so that the order of presentation was balanced. Panelists were provided with appropriate references and paper ballots, and they used a numerical scale from 0 to 15, with one significant digit (Meilgaard et al., 1999). The panelists were also given crackers (Nabisco Premium Unsalted) and water for cleansing and rinsing their palates between each sample.

1.1.2.4. Consumer Testing

Sixty-one individuals were recruited from a database of over 2000 local consumers. Consumers were recruited on the basis of the frequency of their muscadine and muscadine juice consumption. They were also chosen on the basis of their level of liking for muscadine grapes. The 10 juices were tested over the course of 2 days.

A completely randomized serving order was used across the 10 juices for the 61 consumers (Meilgaard et al., 1999). Each consumer saw the first five juices the first day of

Table 1.7. Sensory lexicon used to describe muscadine grape juice.

Term	Definition	Reference	Intensity
Basic tastes			
Sweet	The basic taste sensation on the tongue stimulated by high-potency sweeteners.	Sucrose solutions in spring water.	2.0% solution: 2.0 5.0% solution: 5.0 10.0% solution: 10.0 16.0% solution: 15.0
Sour	The basic taste sensation on the tongue stimulated by acids.	Citric acid solutions in spring water.	0.05% solution: 2.0 0.08% solution: 5.0 0.15% solution: 10.0 0.20% solution: 15.0
Bitter	The basic taste sensation on the tongue stimulated by solutions of caffeine, quinine, and other alkaloids.	Caffeine solutions in spring water.	0.05% solution: 2.0 0.08% solution: 5.0 0.15% solution: 10.0 0.20% solution: 15.0
Aromatics			
Cooked muscadine	The aromatic associated with processed or cooked muscadine grapes.	Post-brand muscadine grape juice	Intensities based on the Universal Aromatic Scale
Cooked grape	The aromatic associated with cooked, processed, or heated grapes but not specifically identifiable.	A mixture of Sunbelt-brand Thompson seedless red grapes cooked and blended together	Same
Musty	The aromatic characteristic of damp/wet basements or turned soil.	2-methyl isoborneol	Same
Green/unripe	Aromatic characteristic of certain green fruits and under ripe fruits in general.	Cis-3-hexenol, Granny Smith apples, underripe fruits	Same
Caramelized/sweet aromatic	The aromatics associated with substances that also have a sweet basic taste such as caramelized sugar, vanillin, maltol, and so on	1) Vanillin, 2) ethyl maltol, 3) caramelized sugar	Same
Floral	A sweet, fragrant aromatic associated with flowers.	Heliotropine	Same
Fermented	Aromatic associated with fermented fruits or grains.	Fermented apple juice, overripe pineapple, cantaloupe, orange juice	Same
Apple/pear	The aromatic associated with different cooked or processed apples or pears.	Gerber-brand apple–pear juice	Same
Metallic	1) The aromatic associated with metal, tin, or iron; 2) the flat chemical feeling factor stimulated on the tongue and teeth.	Dole-brand pineapple juice (canned)	Same
Feeling Factors			
Astringent	The chemical feeling factor on the tongue or other surfaces of the oral cavity described as puckering/dry and associated with tannins or alum.	Alum solutions in spring water.	0.025% solution: 3.0 0.033% solution: 5.0 0.050% solution: 9.0

Universal Aromatic Scale
Intensities for aromatics and aromas are based on the following reference standards: soda note in a saltine cracker = 3.0; cooked apple note in applesauce = 7.0; cooked orange note in orange juice = 10.0; cooked grape note in grape juice = 14.0, cinnamon note in cinnamon gum = 16.0.

testing and the remaining five juices the second day of testing. Consumers received their samples in individual testing booths under controlled lighting and positive airflow. They were provided with a 4-ounce sample of each juice to evaluate overall impression, color, and flavor on a 9-point hedonic scale (9 = "like extremely" and 1 = "dislike extremely"). Aroma, color, overall flavor, muscadine grape flavor, sweetness, and tartness were evaluated on a 5-point Just About Right scale (1 = "much too low," 3 = "just about right," and 5 = "much too high"). Consumers also were asked questions regarding gender, age group, consumption, and purchasing preferences.

1.1.2.5. Summary Results

Table 1.8 provides a summary of the hedonic scores for the 61 consumers. Mean overall liking scores ranged from 6.3 to 7.0 on the 9-point hedonic scale, which is rather small. Without further analysis, it could be concluded that all products were of high quality. However, the use of this data set in Chapters 3 and 6 will clearly demonstrate that segments of consumers with varying preferences exist in this data.

1.1.3. Fried Mozzarella Cheese Stick Appetizers

1.1.3.1. Introduction

This data set is being used in Chapters 8, 9, and 10. Little details are given about the actual samples, as the data are subject to confidentiality agreements. However, the coded data are available for download from http//:www.uark.edu/ua/multivariate/index.htm.

1.1.3.2. Samples

The sample set consisted of eight commercial cheese stick products (P1, P2, P3, P4, P5, P6, P7, and P8). These products were obtained from a food manufacturer and stored in a freezer until evaluation. The samples were fried according to the manufacturers' cooking instructions before serving. On each of two testing days, each product was cooked at each scheduled time according to each product's own cooking instruction using two Red Hots Fryer (Model EF10-120 Deep Fat Fryer, The Eagle Group, Metal Masters Food Service Equipment Co. Inc., Clayton, DE). After cooking, the cooked samples were held under two 250-W infrared bulk lamps (NEMCO Incorporated Food Equipment, Hicksville, Ohio) for no more than 20 minutes. The lamps were used to keep the samples' temperature within the range of 140°–160°F.

1.1.3.3. Consumer Testing

A screening questionnaire was sent via e-mail to potential cheese stick users from a consumer database. Eighty regular cheese stick users (those who ate cheese sticks at least once every 2 weeks) were selected at each of the two testing sites (i.e., the University of Arkansas and Oregon State). A completely randomized design was used across the eight branded samples for all the consumers (Meilgaard et al., 1999). The 2-day consumer tests were carried out at two university sensory laboratories on the same day, using the same computerized ballot. Each consumer was seated in an individual testing booth with controlled lighting and positive airflow and was presented with four samples on each of two consecutive test days. The consumers were served the samples on three-digit coded white plastic plates. Each consumer was asked to evaluate the overall acceptance and acceptance of appearance, flavor, and texture of each sample on a 9-point hedonic scale with 1

Table 1.8. Hedonic and diagnostic means for 10 muscadine grape juices.

Type of juice	Hedonic							JAR	
	Overall	Appearance	Aroma	Color	Flavor	Aroma	Color	Overall flavor	Muscadine flavor	Sweetness	Tartness
Black Beauty	6.9a	6.7cd	6.6abc	6.5cd	6.7a	2.6de	2.7de	2.8d	2.7d	3.3a	2.7d
Carlos	6.8ab	6.7cd	6.7abc	6.5cd	6.7a	2.8bcd	2.4f	3.2ab	3.2a	2.9cde	3.2b
Granny Val	7.0a	6.9cd	6.8abc	6.6cd	6.9a	2.8abc	2.6ef	3.0cd	2.9bcd	3.0cd	3.0c
Ison	6.9a	7.5a	7.0a	7.4a	6.8a	2.7cde	3.0abc	3.1abc	2.8cd	2.9cde	3.3b
Nestitt	6.6ab	7.3ab	6.5bcd	7.2ab	6.5ab	2.7cde	3.1ab	3.1abc	2.8cd	2.8de	3.4ab
Post Red	6.3b	7.3ab	6.4cd	7.3a	6.0b	2.5e	3.1a	3.2ab	2.9bcd	2.7e	3.5a
Post White	6.3b	6.4d	6.1d	6.1d	6.0b	2.5de	3.2a	3.0bcd	3.0abc	2.8cde	3.3b
Southern Home	6.9a	7.0bc	7.0a	6.6cd	6.9a	3.0ab	3.1ab	3.3a	3.2a	3.0bc	3.0c
Summit	7.0a	7.0abc	6.9ab	6.9abc	6.9a	3.0a	2.8cde	3.1abc	3.1ab	3.2ab	2.8cd
Supreme	6.6ab	7.0bc	6.7abc	6.8bc	6.3ab	2.6cde	2.9bcd	3.0cd	2.8cd	2.8de	3.4ab

Note: Means with the different letters for each attribute are significantly different ($p < 0.05$), using the Least Significance Difference method.

Table 1.9. Scales used in the consumer testing of fried mozzarella cheese sticks.

Overall Acceptance and Acceptance of Appearance, Flavor, and Texture	*Amount of Breading*
1 = Dislike extremely	1 = Much too little
2 = Dislike very much	2 = Too little
3 = Dislike moderately	3 = Just about right
4 = Dislike slightly	4 = Too much
5 = Neither dislike nor like	5 = Much too much
6 = Like slightly	
7 = Like moderately	*Saltiness*
8 = Like very much	1 = Much too little
9 = Like extremely	2 = Too little
	3 = Just about right
Purchase Intent	4 = Too much
1 = Definitely would not buy	5 = Much too much
2 = Probably would not buy	
3 = May or may not buy	*Crispness*
4 = Probably would buy	1 = Not nearly crispy enough
5 = Definitely would buy	2 = Not crispy enough
	3 = Just about right
Product Color	4 = Too crispy
1 = Much too light	5 = Much too crispy
2 = Somewhat too light	
3 = Just about right	*Cheese Texture*
4 = Somewhat too dark	1 = Much too soft/melted
5 = Much too dark	2 = Somewhat too soft/melted
	3 = Just about right
Product Size	4 = Somewhat too firm/not melted
1 = Much too small	5 = Much too firm/not melted
2 = Somewhat too small	
3 = Just about right	
4 = Somewhat too large	
5 = Much too large	

= "dislike extremely" and 9 = "like extremely" (Table 1.9). Consumers were also asked to rate the color, the size, the amount of breading, the amount of saltiness, the crispness, and the cheese texture of each sample on a 5-point Just About Right scale, with the Just About Right score = 3 (Table 1.9). Purchase intent toward the products was rated using a 5-point scale with 1 = "definitely would not buy" and 5 = "definitely would buy."

1.1.3.4. Descriptive Analysis

Eight cheese stick products were evaluated for appearance, flavor, and texture by a group of nine Spectrum-trained panelists. Panel orientation was conducted to develop a descriptive lexicon for appearance, flavor, and texture attributes specific to cheese sticks, using the Spectrum method and a numerical scale from 0 to 15 with one significant digit. A total of 44 descriptors (21 flavor attributes, 16 texture attributes, and 7 appearance attributes) were assessed. The list of descriptors is not given here because of confidentiality agreements between the authors and an industrial partner. Texture, flavor, and appearance evaluations were carried out under controlled conditions. During each session, the samples were randomly presented to panelists assigned to an individual booth and provided with a computerized ballot and references. Crackers and water were provided to each

Table 1.10. Consumer hedonic and diagnostic and purchase intent means for eight fried mozzarella cheese sticks.

Sample name	Overall	Hedonic Appearance	Flavor	Texture	Size	Breading	Just About Right Color	Saltiness	Crispness	Texture
P1	6.47a	6.94ab	6.63a	6.11ab	3.13a	3.09a	2.80c	3.09ab	2.63ab	3.61a
P2	6.56a	7.02ab	6.25ab	6.48a	3.03a	3.04a	3.02a	2.86c	2.61ab	3.19b
P3	5.50b	4.99d	5.61c	5.84b	2.06c	2.75b	2.31d	2.84c	2.53abc	3.47a
P4	5.70b	6.13c	5.27cd	5.91b	3.12a	2.88b	2.90bc	2.45e	2.67a	3.02c
P5	4.95c	5.07d	4.97de	5.14c	2.20b	3.04a	2.78c	2.96bc	2.08d	3.11bc
P6	4.31d	4.51e	4.76e	4.11d	1.87d	2.49c	1.81e	2.64d	1.52e	2.56d
P7	6.60a	7.25a	6.22b	6.15ab	3.04a	3.04a	2.99ab	3.13a	2.42c	3.25b
P8	6.27a	6.85b	6.32ab	5.95b	3.07a	3.12a	3.01ab	3.11a	2.49bc	3.54a

Note: Means with the same letter in the column are not significantly different ($p < 0.05$) with the Least Significance Difference method.
JAR = Just About Right scale.

panelist as a means for cleansing and rinsing their palates between each sample. A 10-minute break was scheduled at each session. The texture and flavor evaluations were performed in duplicate, whereas the appearance evaluation, because of product nonuniformity, was performed by presenting several cheese sticks on a white plate.

1.1.3.5. Summary Results

A summary of consumer testing means is provided in Table 1.10 for the eight products tested. Overall liking means ranged from 4.31 to 6.47, with products P1, P2, and P7 being most liked and products P5 and P6 being most disliked. This data set is used in Chapter 5 as an illustration of Landscape Segment Analysis, in Chapter 8 to introduce the concepts of risk analysis, and in Chapter 9 to describe the application of multivariate adaptive regression splines to identifying attributes driving liking of this type of product.

1.1.4. Data Sets for Panelist and Panel Performance Evaluation

1.1.4.1. Chocolate Chip Cookies—Simple DA

This experiment dates back to the 1980s. Chocolate chip cookies are a common product across North America. There are clear differences among commercial products that result from the use of real chocolate chips versus compound chips and the proportion of chips in the formula. In recent years, chewy biscuits have been included in this product category. It is easy to find four or five brands that are sufficiently different that a novice panel may be trained in the evaluation of a small number of key attributes in a short period of time, usually under 1 hour. This data set was created in 2002 as part of a training exercise. The attributes were familiar to most potential panelists. In this case, the Excel (Microsoft Corp., Redmond, WA) sheet held responses from 10 panelists for five attributes and five products in a single session. Because this was a training session, the samples were presented monadically in a fixed order.

1.1.4.2. White Wine Descriptive Panel

This data set was part of a major study that has been reported in the literature (Findlay et al., 2006). A total of 20 white wines were evaluated in triplicate by 10 panelists using 76 attributes. For this reduced data set, there are 10 wines, and the 27 attributes are those related to flavor and mouthfeel. The data were collected over nine sessions that provided three replicates. The design was a balanced complete block that was broken across sessions.

References

Anonymous. 1998. An interview with McCormick on trends in the snack food industry. Cereal Food World 43(2):60–65.
Brooks, R.M. and H.P. Olmo. 1991. Register of new fruit and nut varieties. List 36. HortScience 26:964–967.
Brooks, R.M. and H.P. Olmo. 1970. Register of new fruit and nut varieties. List 25. HortScience 5:384–385.
Buttery R.G. and L.C. Ling. 1995. Volatile flavor components of corn tortillas and related products. J Agri Food Chem 43: 1878–1882.
Buttery R.G. and L.C. Ling. 1998. Additional studies on flavor components of corn tortilla chips. J Agri Food Chem 46:2764–2769.
Dies C.R. 2000. Food Product Design: Applications. Available at http://www.Foodproductdesign.com/archive/2000/0100ap.html. Accessed on December 2, 2000.

Ector, B.J. 2001. Compositional and nutritional characteristics p. 341–367. In: Muscadine Grapes. F.M. Basiouny and D.G. Himelrick (eds.). Alexandria, VA: ASHS Press.

Ector, B.J., J.B. Magee, C.P. Hegwood, and M.J. Coign. 1996. Resveratrol concentration in muscadine berries, juice, pomace, purees, seeds, and wines. Am. J. Enol. Vitic. 47:57–62.

Findlay, C., J. Castura, P. Schlish, and I. Lesschaeve. 2006. Use of feedback calibration to reduce the training time for wine panels. Food Qual. Pref. 17:266–276.

Goldy, R.G. and W.B. Nesbitt. 1985. 'Nesbitt' muscadine grape. HortScience 20:777.

Green, C. 2003. Nutraceuticals and pharmaceuticals—market is growing rapidly for medicinal foods. Ag. Innovation News 12(1).

Hawrysh Z.J., M.K. Erin, S.S. Kim, and R.T. Hardin. 1995. Sensory and chemical stability of tortilla chips fried in canola oil, corn oil, and partially hydrogenated soybean oil. J. Am. Oil Chemist Soc. 72:1123–1130.

Lane, R.P. 1977. 'Summit' muscadine grape. HortScience 12(6):588.

Lanier, M.R. and J.R. Morris. 1979. Evaluation of density separation for defining fruit maturities and maturation rates of once-over harvested muscadine grapes. J. Am. Soc. Hort. Sci. 104:249–252.

Lee, J.H. and S.T. Talcott. 2004. Fruit maturity and juice extraction influences ellagic acid derivatives and other antioxidant polyphenolics in muscadine grapes. J. Agric. Food Chem. 52:361–366.

Lisser, E. 1993. Tortilla chips tempt snackers with changes. Wall Street Journal, May 6, p B1.

Meilgaard M., G.V. Civille, and B.T. Carr. 1999. Sensory Evaluation Techniques. 3rd Ed. Boca Raton, FL: CRC.

Meullenet, J.-F., R. Xiong, M. Monsoor, T. Bellman-Horner, S. Zivanovic, P. Dias, H. Fromm, and Z. Liu. 2002 Preference mapping of commercial toasted white corn tortilla chips. J. Food Sci. 67:1950–1957.

Meullenet, J-F., R. Xiong, J.A. Hankins, P. Dias, S. Zivanovic, M.A. Monsoor, T. Bellman-Horner, Z. Liu, and H. Fromm. 2003. Modeling preference of commercial toasted white corn tortilla chips using proportional odds models. Food Q. Pref. 14(2003):603–614.

Mortensen, J.A., J.W. Harris, and D.L. Hopkins. 1994. 'Southern Home': an interspecific hybrid grape with ornamental value. HortScience 29:1371–1372.

Noguera, E., Morris, J., Striegler, K., and Thomsen, M. 2005. Production budgets for Arkansas wine and juice grapes. Ark. Agri. Expt. Sta. In press.

Pastrana-Bonilla, E., C.C. Akoh, S. Sellappan, and G. Krewer. 2003. Phenolic content and antioxidant capacity of muscadine grapes. J. Agric. Food Chem. 51:5497–5503.

Rea L.M. and R.A. Parker. 1992. Designing and conducting survey research: a comprehensive guide. San Francisco: Jossey-Bass.

Stinson, C.T., and Tomassetti, S.J. 1995. Effect of natural corn flavor on the sensory quality of low fat tortilla chips. Food Tech. 49(5):84, 88–90.

Striegler, R.K., P.M. Carter, J.R. Morris, J.R. Clark, R.T. Threlfall, and L.R. Howard. 2005. Yield, quality and nutraceutical potential of selected muscadine cultivars grown in southwestern Arkansas. HortTechnology 15(2):276–284.

Talcott, S.T. and J.H. Lee. 2002. Ellagic acid and flavonoid antioxidant content of muscadine wine and juice. J. Agric. Food Chem. 50:3186–3192.

Threlfall, R.T., J.R. Morris, L.R. Howard, C.R. Brownmiller, and T.L. Walker. 2005. Pressing effects on yield, quality, and nutraceutical content of juice, seeds, and skins from Black Beauty and Sunbelt grapes. J. Food Sci. 70(3):167–171.

Wellman D. 1997. Snack foods. Supermarket Bus. 52(9):64.

Yilmaz, Y. and R.T. Toledo. 2004. Major flavonoids in grape seeds and skins: antioxidant capacity of catechin, epicatechin, and gallic acid. J. Agric. Food Chem. 52:255–260.

2 Panelist and Panel Performance: A Multivariate Experience

2.1. The Multivariate Nature of Sensory Evaluation

Sensory evaluation is a multivariate task. When we ask individuals to evaluate the characteristics of a product, they have to make that evaluation within the context of the product. When we select and train people to become members of a sensory panel, we attempt to determine their innate abilities, and then we extend their sensory knowledge through training exercises. In many cases, the screening and training are conducted using pure compounds as individual stimuli that are presented and evaluated in isolation. It is quite easy to judge the saltiness of a solution of sodium chloride in distilled water. The task becomes significantly more difficult if saltiness is being determined in a salsa that contains sweet, sour, bitter, and chili heat, in combination with other seasonings, tomatoes, and other vegetables.

As a result, when we consider the performance of individual panel members or that of the panel as a whole, we must use techniques that measure performance in context. The measurement of panel and panelist performance is essential to the determination of proficiency. Accuracy and precision are fundamental measures of analytical values. These values are difficult to judge in the absence of absolute or universal standards. Nevertheless, we must attempt to evaluate these elements of proficiency if we are to judge the progress of training descriptive sensory panels and determine the confidence of repeatability of panel results over time. For example, it would be useless to perform a shelf-life study if the descriptive analysis at each time point was entirely unrelated to the previous one. In the absence of sensory absolutes, we must work hard to calibrate our panels with reliable references and consistent sensory protocols. Although univariate measures are very useful, the multivariate analysis of descriptive data provides a more comprehensive picture. In this chapter, two sample data sets will be used to provide examples. Full reference to these descriptive studies may be found in Chapter 1.

2.1.1. A Musical Analogy

The analogy of a symphony orchestra can be used to create a basis for understanding both the performance of an individual panelist and that of the entire panel as a group. To begin, we can draw a connection between the sensory properties of a product and the sounds produced by an orchestra. Each musical instrument produces a unique sensory signature, which we can think of as an attribute. Some instruments can only produce high notes, whereas others are limited to the bass range. Some can play very loud, and others are only soft.

When we join the audience for a concert, we become members of a sensory panel. Each one of us will have inherent strengths and weaknesses in our individual perception. The other contributing factor is the extent of our experience and musical training. Musicians practice extensively to become familiar with a musical piece. They must be able to play without thinking about every note before they play it. This can be easily observed by listening to a person practicing their first, easy, pieces on a piano. Initially, they are unfamiliar and hesitate as they progress through the music. Over time, they can produce what appears to be an effortless performance. This analogy holds particularly true for descriptive sensory panelists. Initially, they may have never experienced a product or the set of sensations that it may produce. In the absence of that prior experience, they have no vocabulary to describe new sensations. When they do manage to describe the sensation, it may be using terms that are only meaningful to themselves and difficult to communicate to others. Training provides a wonderful opportunity to expose panelists to new sensations and to provide the common labels that can be used to communicate those sensations to others. Hildegarde Heymann has observed that the limitations of the young or the illiterate in carrying out descriptive analysis tasks do not exist at the sensory level. They are caused by lack of exposure to the products and to the absence of a vocabulary to express their experience. It takes longer to train these groups because they must first be educated to enlarge their sensory exposure and their vocabulary (Lawless and Heymann, 1998). This may be thought of as a creation of context. It has been said that the Inuit have many words for snow but few words for sand. In contrast, the Bedouin have many words for sand but few for snow. A descriptive sensory term can only become part of a useful lexicon if it has a clear definition and has an example that can be communicated to anyone, regardless of their personal experience. By having a reference material, we can use a word like "skunky" without ever having encountered a live skunk.

Understanding the language of descriptive sensory analysis is similar to learning how to listen to music. There is a theoretical component of understanding the basis of a term, and then there's the practical aspect of being able to recognize the sensation in products. It is relatively easy to identify a lone piccolo and describe the notes that it plays. It is entirely different when an entire orchestra plays together. This is the same sensation that we have on exposure to a complex sensory product for the first time. All the sensations run together, and we are overwhelmed and can only respond to it as "noise." However, as we have been previously trained to identify the sounds of a piccolo, we can work to isolate that instrument from the other instruments in the orchestra. To take this analogy a bit further, when the sound of the piccolo is soft, it may be impossible to pick it out from the background of sound. However, when the rest of the orchestra plays softly, the piccolo sound is very clear. There are some sensations that are like the timpani, which can blast its way through, regardless of the rest of the orchestra. Intense basic tastes have the ability to dominate the sensory response to a food. For example, an extremely bitter product can overcome any of the nuances that may exist in a product. This is the principle put forward by Pascal Schlich and his colleagues in the Temporal Dominance of Sensations method (Pineau et al., 2004).

To come back to the orchestral analogy, when a soloist takes the lead in a musical piece, the remainder of the orchestra continues to play a supporting role, but our attention is focused on the solo. The sensation we experience is not felt in isolation—it appears against a background that modulates our response. This is an important clue to the mystery of understanding the difficulties of sensory evaluation of even well-defined attributes in complex product systems.

2.1.2. Understanding Attributes in Context

The profound effect of context on sensory perception means that to train panelists to be effective in the profiling of complex products, we must follow an order-of-operations approach that relies first on unambiguous identification of the attribute, second on the ability to rank intensities of that attribute against the background of the product, and finally, scaling of the attribute, providing the attribute and the context permit discrimination. We do this in panel training by providing feedback in a variety of forms: verbal, group, individual, written, and numeric.

The key to making feedback work is providing "true" information to the panelists in a timely fashion. This permits calibration of the response. However, if the feedback is either trivial or incorrect, the panelist will be confused and the desired learning will not take place. To establish meaningful targets for feedback, it is important to understand the shape of the psychometric function and the portion of the curve that describes the attribute intensity for the product being studied. Although the same attribute may be identified and defined in a range of different products, its perception will be dependent on the product being tested. In simplest terms, sucrose perception will behave quite differently in a citrus drink than it will in water. Both the Just Noticeable Difference and the threshold values will be influenced by the other components and stimuli in any system. Attributes can be assigned to several categories, reflecting their degree of sensory difficulty, which can assist in applying the most appropriate strategy for both training and the collection of data. By understanding the dynamics of attributes, it is possible for sensory analysts and panel leaders to refine the process of training panels.

Analysis of the data sets used in this chapter was performed using Senstools Version 3.3.1 (2005).

2.2. Univariate Approaches to Panelist Assessment

2.2.1. Visualizing Raw Data

Every time I talk to sensory students about data analysis, I start by saying, "Don't be afraid to look at the raw data." However, it takes a lot of time to review each data point in a descriptive study of 10 products by 10 panelists for 50 attributes and 3 replicates (15,000 points). To accelerate this process it is much easier to review data in graphic form as scatter plots. The human eye and brain have an excellent capacity for pattern recognition. However, we can be fooled into seeing things that are not really there, so caution is required. If we prepared a plot of scores by panelist for each attribute for each product including replicates, we would have 500 graphs to view. An approach that simplifies this is to arrange all of the plots in the same display. Although many researchers have used this technique, it was Frank Rossi (2001) who presented this approach at the 6th Sensometrics Meeting.

Scatter plots provide a quick overview of sensory data by graphing the raw values for each panelist by product and attribute. The scatter plots provide a tool for rapid identification of panelists, products, and attributes that are very different. Panelist-oriented scatter plots (Figure 2.1) allow an analyst to look at the panelists' raw data by product and by attribute. Large scatter tells us that the panel is in poor agreement and that the attribute may be a problem. Because we can review all panelists at the same time, we can determine whether the problem is specific to one individual or whether all panelists have the same problem for the same attribute. We can also rapidly identify outlying assessments and see whether they are anomalous. The product orientation of the scatter plot (Figure 2.2) allows us to see which panelist and his or her replicate is out of agreement with the rest of the panel. A separate plot is created for each product/attribute combination. Each column displays the plots for one attribute, with the value range noted at the top and bottom of

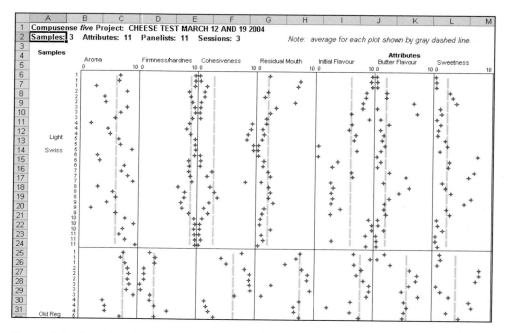

Figure 2.1. Example of scatter plots of panelist-oriented data for individual attributes and replications.

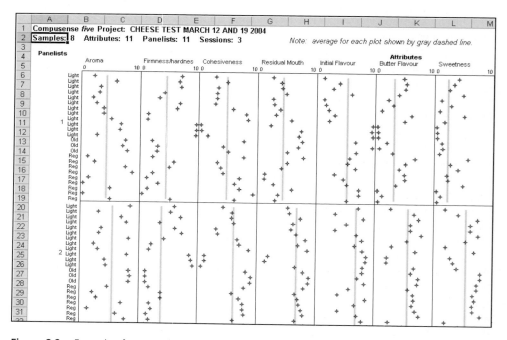

Figure 2.2. Example of scatter plots of product-oriented data for individual attributes and replications.

the column. Each row displays the plots for one product. Panelists are identified by their codes at the left. The gray line appearing within each plot is the average for all the values. Further investigation may then be done with more sophisticated statistical methods.

2.2.2. PanelCheck

The sensory science group at the Norwegian Food Research Institute, Matforsk, has been at the forefront of statistical research into panelist performance. Matforsk has been conducting descriptive analysis projects for more than 30 years and has constantly monitored the performance of both their panel and its members. Under the leadership of Einar Risvik, original research has been conducted and published by many Matforsk scientists, including Tormod Naes, Per Lea, and Marit Rodboten.

The topic of dealing with individual differences between panelists in descriptive analysis has been the subject of extensive work by Tormod Naes (Naes, 1998). He documents a variety of attempts to normalize scale data mathematically and provides a cautionary discussion on the pitfalls. It is inevitable that there will be a scaling effect; this is recognized as a part of the generalized procrustes analysis (GPA) and results in weighting values that are applied to panelist scores to harmonize them with panel mean values. This source of variance in sensory data is normal and must be taken into account in evaluating panelist performance. The best strategy to deal with this is through panel calibration with suitable reference standards. Naes recommends a nonstructured scale with calibration of the upper and lower end of the scale (Naes, 1990).

A major initiative undertaken at Matforsk has been the development of a series of statistical tools written in MATLAB to perform a wide range of tests, mainly based on analysis of variance (ANOVA), that help panel leaders understand panel performance. This package, known as PanelCheck, has been used extensively in the Scandinavian countries. Asgeir Nilsen and colleagues (2004) presented the most recent version of this tool at the Seventh Sensometrics Meeting in Davis, California, in 2004. It provides several unique methods of analysis; most are univariate, but there are several multivariate approaches. One of the most interesting is the Eggshell Plot. This approach permits the comparison of a panelist's actual assessment of intensity of an attribute over a wide range of intensity and across products with the ideal or panel consensus value. The plot provides a parabola that resembles the surface of an egg, and the values for the panelist present the broken shell. The closer the panelist is to the ideal value, the smoother the curve. Large departures from the ideal indicate poor performance at that attribute intensity. This addresses the phenomena whereby a panelist performs well at near-threshold levels of an attribute but poorly at higher concentrations of the same stimuli. This approach also can be used to demonstrate the reverse of this situation, when the panelist is poor at threshold but good at high concentrations. This type of analysis requires relatively large quantities of different measurements to provide a clear picture.

Another approach to performance of a panelist is to plot mean square error versus p value for all attributes by one assessor. This gives a graphic example of repeatability and discrimination. Problem attributes can be identified rapidly. This approach can be used to plot all panelists for a single attribute, and even all assessors and all attributes. The quantity of data can make the plots somewhat difficult to read.

The PanelCheck tool is under further development and revision by a team lead by Margrethe Hersleth at Matforsk, so it will be worth watching for the next stage in its development.

2.2.3. Feedback Calibration

The Feedback Calibration method is a powerful tool for training an optimum sensory descriptive analysis panel. The technique is based on original research conducted by Compusense over the last 3 years. It uses the application of immediate feedback as a tool to reinforce learning and to provide calibration during the sensory task. The research was conducted in two phases. The initial project used two red wine panels and a series of 20 wines that had been previously profiled by a conventionally trained panel. One panel received only computerized feedback, whereas the other received only conventional debriefing after the sessions. Both panels were found to be equal to the conventional panel in proficiency. This research was originally presented at the 2003 Pangborn Symposium in Boston. The second project used two white wine panels, one made up of previously trained red wine panelists and the other comprising complete novices. Both were oriented to wine aromas using the Wine Aroma Wheel, developed by Ann Noble. During training, the panels developed their own feedback targets that were refined and calibrated during training. At the end of nine sessions of training, each panel evaluated 20 wines in triplicate. Validation of the results of the panels was conducted with the method, developed by Pascal Schlich, using the normalized deviation between the observed regression vector (RV) and the calculated RV (the NRV). The panels were shown to have been able to train themselves to proficiency in half the time it takes a conventional panel (Findlay et al., 2006).

2.3. Multivariate Techniques for Panelist Performance

The power of multivariate analysis is in its ability to simplify the presentation of complex data. All methods have a common goal: to provide a graphical output that permits the analyst to understand the differences and similarities among products. This approach may be applied to the performance of panelists and panels.

2.3.1. Multivariate Analysis of Variance

Examining the difference between univariate ANOVA and multivariate analysis of variance (MANOVA) requires us to think in terms of products being located in a sensory space, rather than single-attribute means for each product. The product locations within that sensory space become expressed as matrices. For a complete background on this approach, refer to Schlich et al. (2004).

The concept of dimensionality is essential to understanding the relationship between the sensory properties of products in a sensory space. The large numbers of attributes that are frequently generated during descriptive analysis are reduced in multivariate analysis to the dimensions used for plotting in principal components analysis (PCA) or GPA. This simplification of the data allows us to visualize complex relationships; however, we lose information as a result of the collapsing of attributes to dimensions. Many of the attributes are, in effect, multicollinear. In their product environment, they are related and vary in some linear relationship.

For example, it is easy to see in the aging of spirits in wooden barrels that the attributes associated with the wood all increase together. If there is one indicator attribute that captures the contribution of that component, it may be sufficient to differentiate products that are different.

The number of principal axes in a multivariate model describes the complexity of the sensory information being obtained. It is presumed that a larger number of dimensions is better.

Pascal Schlich has created a model of measurement of performance of both panelist and panel in both univariate and multivariate approaches (Schlich et al., 2004) For multivariate analysis of panelist performance, he recommends MANOVA as a test of product discrimination, canonical variate analysis as a way to determine dimensionality of product configuration in the sensory space, and a step-wise selection of the most important attributes that describe product differences. For the panel, he again uses the MANOVA test but measures homogeneity with the product × panel interaction. Discrimination may be determined using the MANOVA test of product in a model of product and panelist and product × panelist interaction. Canonical variate analysis is used for dimensionality, and confidence intervals on the map permit estimates of homogeneity and discrimination.

A major contribution to the valid comparison of panel homogeneity was made by Schlich with the application of several matrix methods including RV (vector correlation), alpha coefficient and RVD to panel data. RV measures product configuration, alpha compares product scores, and RVD evaluates the attribute correlations. These methods all produce values that range between 0 (no similarity) and 1 (identical). A method was developed to normalize these values so that confidence measures could be applied to the results. Details of these methods may be found in Schlich et al. (2004). If, in a study, there are attributes that can be divided into appearance, aroma, flavor, and texture, this study becomes easier to compare to another study with the same modalities, even when the second study has different attributes. Sensory modalities may be applied to facilitate the comparison of different studies.

2.3.2. Normalized RV

It is possible to calculate the degree of similarity between two data sets or configurations by using the RV coefficient. RV can be understood as a correlation coefficient in a multidimensional space ($0 < RV < 1$)—the closer RV is to 1, the more similar the configurations of the two sensory spaces are. By conducting a large number of random and independent permutations, it is possible to draw the distribution of the RV calculated this way. Using a normal distribution (N), a Z can be calculated and a nonparametric test statistic created. Kazi-Aoual et al. (1995) gave the analytical expression of the mean and variance of the N-RV coefficients. This expression was calculated by the permutation of n lines of one of the data sets and by comparison to the other data set by calculating RV. A normalized deviation between the observed RV and the calculated RV, called the NRV, can be computed. If NRV is greater than 2, the similarity between the two tables can be considered to be higher than what would be calculated by chance. Data from two panels can be used to validate one another's sensory spaces; for example, the NRV based on permutation tests was calculated to determine whether the RV calculated was different than an RV generated by chance (Schlich, 1996). An NRV larger than 2 indicates that two configurations are significantly ($p = 0.05$) more similar than when product labels are permuted within one configuration. An NRV between 1.5 and 2 indicates that two configurations are slightly similar, and an NRV between 1 and 1.5 indicates almost no similarity.

The NRV values in Table 2.1 provide an excellent tool for comparing the performance of two panels that profiled the same products, using different attributes. The four sensory modalities tested show great overlap for panel T. The attributes assessed for aroma before stirring, aroma after stirring, and flavor are all significant. In essence, the same attributes

Table 2.1. Normalized regression vector coefficients for Panel T and Panel U for aroma before stirring, aroma after stirring, flavor, and taste/mouthfeel.

		Panel T				Panel U			
		ABS	AAS	FLA	TMF	ABS	AAS	FLA	TMF
Panel T	ABS	—							
	AAS	9.5	—						
	FLA	3.8	4.6	—					
	TMF	2.1	2.1	10.9	—				
Panel U	ABS	**1.8**	3.5	1.5	0.2	—			
	AAS	0.4	**1.3**	1.4	0.5	1.8	—		
	FLA	2.7	2.8	**11.4**	11.6	1.4	1.3	—	
	TMF	2.4	2.4	10.9	**11.8**	0.5	1.0	12.8	—

Note: Bold figures are direct comparisons. ABS = aroma before stirring, AAS = aroma after stirring, FLA = flavor, TMF = taste/mouthfeel.

are being measured three times over. In panel U, there is much less overlap, and about half as many attributes were used, as they were not repeated across modalities. When we look at the relationship between the panels, both flavor and taste/mouthfeel are highly correlated, showing that the panels were in excellent agreement about the sensory properties of the products tested.

2.3.3. Principal Component Analysis

PCA is discussed in detail in other parts of this book. Two examples of the application of PCA to panelist and panel performance will serve to illustrate the usefulness of this multivariate method. The data set for chocolate chip cookies has been analyzed to provide the plots for the first example.

The PCA plot (Figure 2.3) shows the product sensory space that separates each of the five samples. Four of the attributes are significant ($p < 0.05$), and two of them, chocolate flavor and quantity of chocolate chips, are collinear. The two dimensions accounted for 92% of the variance. In a simple system like this, where the products are quite different in intensity for each attribute, this high variance accounted for is not unusual. If we conduct PCAs by attribute for products and panelists and plot them together, it is easy to see the panelists who are in good agreement. In Figure 2.4, for quantity of chocolate chips, we know that sample E has the greatest number; however, panelists 6 and 2 have judged the opposite from the panel. There are two possibilities: either they cannot count, or the samples they received were anomalous. Because this result can be obtained immediately during training, it is easy to determine whether the problem was with the sample or with the panelist. In either case, it can be resolved quickly.

Hardness is a more difficult case (Figure 2.5) because samples A, C, and E are not significantly different. However, the majority of the members of the panel are in good agreement, as they cluster at the left end of the first dimension. Panelists 5 and 6 are not in agreement with the panel. This could be a case of scale reversal, which can be dealt with in the debrief session. If we look at the plot for sweetness (Figure 2.6), the one attribute that was not significantly different among samples, there is a dispersion of both products and panelists, with no clear direction. This is the reason that we only give feedback to the panel if the result is unambiguous.

The second application of PCA is in the comparison of the sensory spaces described by the two wine panels (Figure 2.7). In this case, panels U and T profiled the same 20 white

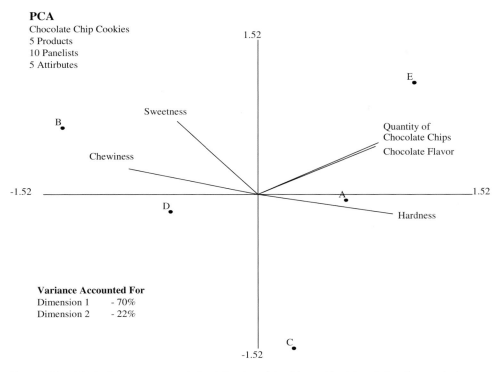

Figure 2.3. Principal component analysis of the chocolate chip cookie data set showing product scores and variable loadings.

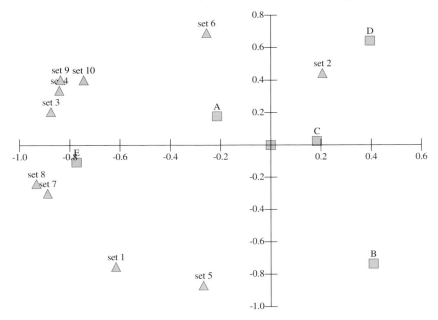

Figure 2.4. Principal component analysis of panel data for the attribute quantity of chocolate chips.

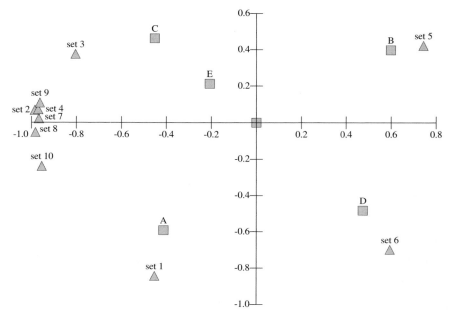

Figure 2.5. Principal component analysis of panel data for the attribute hardness.

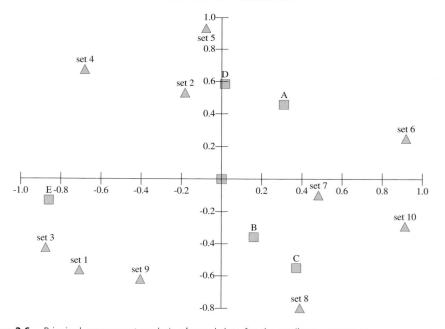

Figure 2.6. Principal component analysis of panel data for the attribute sweetness.

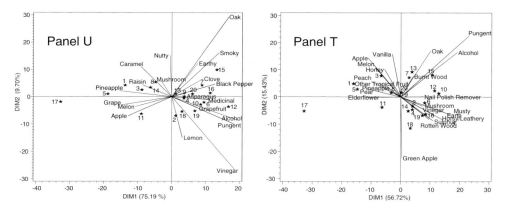

Figure 2.7. Principal component analyses of sensory data from two wine panels showing product scores and variable loadings.

wines (Findlay et al., 2006). Although the two panels used different attributes (panel U used 20, and panel T used 22), the PCAs for flavor are remarkably similar in their positioning of products in the two principal sensory dimensions. Variances accounted for by panels U and T were 85% and 72%, respectively. As indicated in Table 2.1, the relationship between these sets of data showed an NRV of 11.4, which is highly significant, supporting the graphic comparison from the PCA.

2.3.4. Generalized Procrustes Analysis

The origin of GPA has a strong basis in the comparison of panel results across languages. It was developed to permit statistical analysis of the results of Free Choice Profiling, which allows individuals to develop their own terms for attributes that describe a range of products in their own words or language. It is the only way to make sense of Free Choice Profiling data. The central concept is that all panelists see the same products, so the products do not vary. As a consequence, a mathematical consensus may be built that uses multidimensional rotation, translation, and scaling to account for panelist differences. Once established, that information is used to analyze the differences among products. GPA uses individual responses rather than the mean values typically used in PCA. This gives a measure of variance in the analysis.

To illustrate the value of GPA in panelist and panel assessment, a complex product data set from white wine evaluation will be used. The GPA plot for the first two dimensions may be seen in Figure 2.8. Most of the wines are well separated and cover the sensory space well. Of the 27 attributes, all but four were found to be significant ($p < 0.05$). Because GPA estimates the scaling factor applied to panelist scale usage, we have a weighting factor that is used to compensate for individual usage. Figure 2.9 shows that although we have very good discrimination of products, some panelists—P2, P5, and P6—use the lower end of the scale (high weights), and others—like P9—use the upper end of the scale. There are several ways we can examine panelist performance graphically. The assessor plot is a GPA that visualizes assessors on the basis of all products, attributes, and replicates (Figure 2.10). Interpretation of the plot is not entirely clear, but great dispersion or polarization provides evidence of panelists who are in poor agreement. This evidence can also be examined through panelist correlations to the consensus. Figure 2.11 shows results for

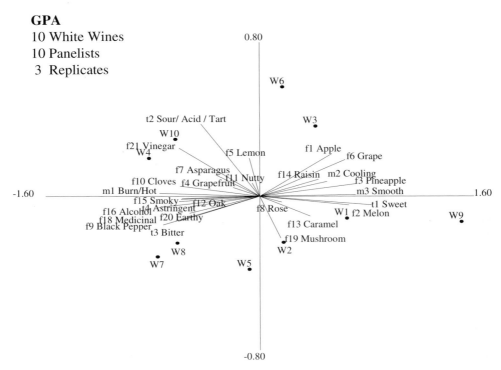

Figure 2.8. Generalized procrustes analysis of sensory data representing 10 white wines (W1–W10), 10 panelists and three replicates.

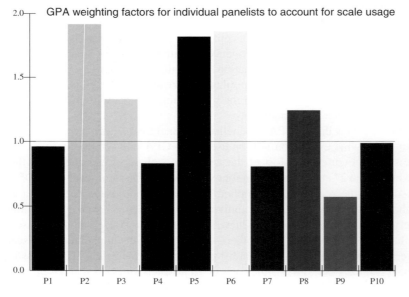

Figure 2.9. Panelists' weighting factors accounting for scale usage generated from generalized procrustes analysis.

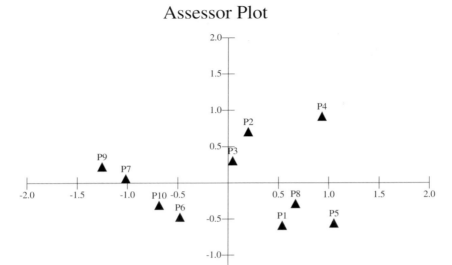

Figure 2.10. Generalized procrustes analysis panelist plot (P1–P10 represent panelists).

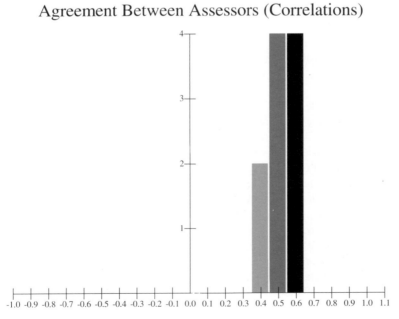

Figure 2.11. Panelists' data correlations for the white wine study (x-axis = Pearson's correlations, y-axis = number of assessors).

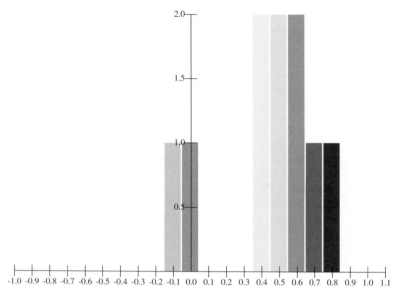

Figure 2.12. Panelists' data correlations for the chocolate chip cookie study (x-axis = Pearson's correlations, y-axis = number of assessors).

the white wine study, where the correlation of four panelists with the consensus is at 0.6 and 0.5, respectively, and two panelists are at 0.4. All of these are positive correlations. Figure 2.12 shows the correlations for the chocolate chip cookie panel, ranging from −0.1 to 0.8, illustrating poor overall agreement.

Another way of judging our panelists is by how well each panelist uses the GPA sensory dimensions to describe the products. The screeplot of variance accounted for by dimensions (Figure 2.13) for each panelist follows much the same pattern for all panelists. This confirms their ability to use the attributes to discriminate between products in a similar fashion.

The ability to provide individual vectors for panelists is one of the most useful aspects of GPA. In training, the rapid feedback of this information to panelists is invaluable in creating consensus. PCA by attribute provides similar information that can lead to the same conclusion, but it is not as elegant or easy to read as the vector plots on the GPA. Figure 2.14 shows the PCA and Figure 2.15 the GPA for the alcohol attribute in the white wine study. The message from both of these plots is the same: The panel appears to be in agreement. However, in the case of P8 using PCA, we would wonder whether he or she is in line with the panel. The GPA indicates that all panelists are in excellent agreement and are discriminating between the products well.

If we look at the GPA for asparagus (Figure 2.16), an attribute that was found by ANOVA to be significantly different across products ($p = 0.01$), we can clearly see that panelist 5 is at complete odds with the remainder of the panel. It may well be that this panelist suffers from either anosmia or the misidentification of the attribute. The panel leader now has the information to determine what the problem may be. In the case of the Lemon attribute (Figure 2.17), the ANOVA is not significant, and there is no agreement among the panelists.

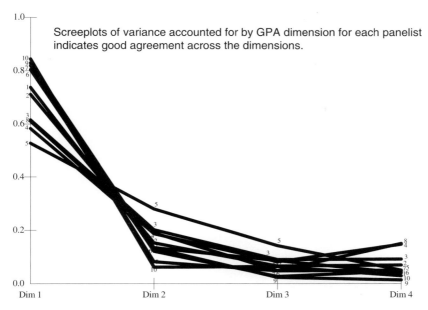

Figure 2.13. Screeplot of variance accounted for by the first four GPA dimensions for each panelist.

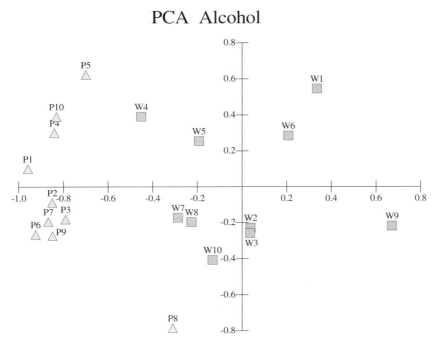

Figures 2.14. Principal component analysis showing product scores and variable loadings (panelists) for the attribute alcohol (W1–W10 are wine products, P1–P10 represent individual panelists).

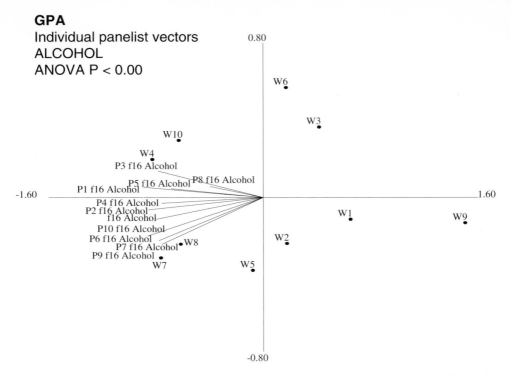

Figure 2.15. Generalized procrustes analysis showing product scores and panelist vectors for the attribute alcohol in the first two dimensions (W1–W10 are wine products, P1–P10 represent individual panelists).

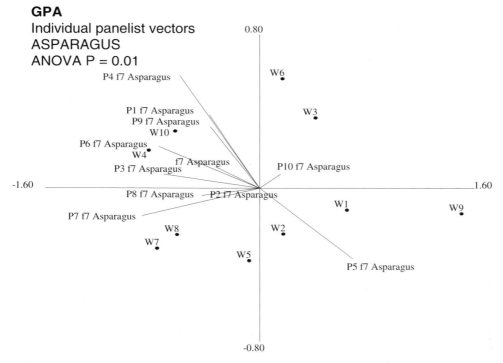

Figure 2.16. Generalized procrustes analysis showing product scores and panelist vectors for the attribute asparagus in the first two dimensions (W1–W10 are wine products, P1–P10 represent individual panelists).

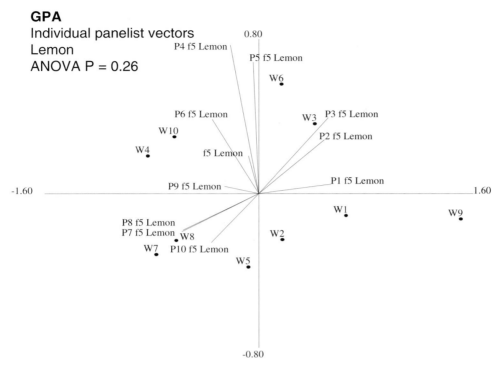

Figure 2.17. Generalized procrustes analysis showing product scores and panelist vectors for the attribute lemon in the first two dimensions (W1–W10 are wine products, P1–P10 represent individual panelists).

GPA may also be used as a method to evaluate different panels. Figure 2.18 shows three panels that were all part of the same study (Findlay et al., 2006). The original panel, identified as D, was trained and then profiled 20 red wines. Following this, two new panels were recruited and trained over a 3-week period. The progress of their training every 3 days is mapped in this plot. The control panel, C, and the experimental panel, E, may be tracked moving from left to right across the plot. Over the course of training, these panels are shown to move toward each other and closer to the D panel. In this case, the matrix of data for the test was rotated to treat the panels as if they were products and the products as if they were panelists. It is of interest to note that the three replicates of the D panel are virtually superimposed, indicating excellent replication.

2.4. Panel Evaluation through Multivariate Techniques

2.4.1. GPA for Lexicon Reduction

Byrne and his colleagues (2001) at KVL in Copenhagen were concerned about the consistency of panels in the development of a lexicon of terms to describe a warmed-over flavor in pork. The researchers systematically reduced the attributes from 42 terms to 20 terms over seven sessions. GPA was used to evaluate the effect of each reduction by using the variance accounted for by dimension and by session. The researchers showed that after

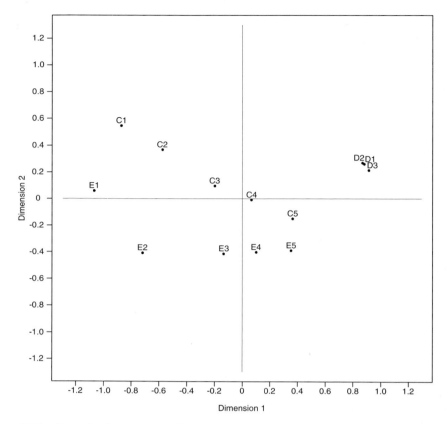

Figure 2.18. Generalized procrustes analysis comparing replicated assessment performed by three different red wine panels (D = original panel; panels C and E being analyzed against D at specific intervals during training).

five sessions, the variance accounted for in the first dimension increased from the 35% level to the 50% level. The other dimensions appeared to be quite stable. The reduction in the number of attributes being measured was justified. It is also interesting to note that the authors concluded that warmed-over-flavor is "not a unidimensional phenomenon but rather a multidimensional phenomenon from a sensory perspective" (p. 187).

Clearly the performance of a panel depends on the comprehension of the set of terms being used: With the comprehension of too few terms, the panel members will not capture the sensory space of the product; too many, and they spend energy on the research but provide no greater information or value. A caution that is frequently forgotten is that attributes that do not discriminate within a particular set of samples may become significant when new products are introduced. This is the reason that panelists are encouraged, and even compelled, to report any new attributes that they notice during sensory evaluation.

2.4.2. The European Sensory Network Multicountry Projects

The European Sensory Network conducted a five-panel multicountry descriptive analysis study of coffee, conducted under strict protocols. The data set from this work has been a veritable gold mine for further statistical analysis. Pascal Schlich (1998) applied his FLASH analysis, which examined attribute consistency over panels, as well as the attributes that were specific to a panel. This approach revealed that of 16 attributes, seven demonstrated significant ($p < 0.05$) sample effects; however, significant sample × panel interaction occurred in all but three attributes. Of the significant attributes, only two showed $p > 0.05$ by panel; however, rancid flavor, which is not significant in the pooled model, is significant when panel is a factor. Schlich introduced the observation that meta-analysis assumes that when a small, nonsignificant effect is observed systematically over different studies, it should be viewed as significant. Of course, this leads us back to the question of the validity of the results of the descriptive analysis. Of the five panels, one was unable to declare a significant difference across the eight coffees in the study. Of the panels that declared differences, they numbered eight, four, three, and two attributes, respectively, that were significant. Regardless of this disparity, each panel's results led to the same conclusion in the categorization of the coffees by sensory description.

The European Sensory Network also conducted a 12-panel multicountry study of red wine to investigate panel proficiency under the European Union ProfiSens project (McEwan et al., 2002). This research highlighted the many difficulties encountered in conducting sensory evaluation on complex products across panels that vary greatly in their product experience and that are almost impossible to standardize. An observation that was particularly important was that the differences among the samples being used in the proficiency test must be well established in advance. This helped in the planning for the design of Findlay et al.'s (2006) feedback calibration study, which relied on a well-trained and experienced red wine panel to establish the profiles of the 20 test wines used in the experimental part of that research. The groups that participated in ProfiSens each trained their own panels independently. When a common set of attributes was examined, there was poor correlation among most of the panels. In the absence of specific training references, it is virtually impossible to make this approach work. A scheme composed of five steps was followed to assemble an overall score that reflected the sensory dimensionality, discrimination, agreement with expected sensory map, and assessor agreement within and across panels. In the case of the comparison of panel performance for panel proficiency, an expected result score had been established in advance, and only two validation panels exceeded the target for proficiency. As a consequence, panel comparisons were made using the RV approach, which has already been described (Schlich, 1998). In the absence of a consensus of all panels, an RV exceeding 0.7 is considered desirable. The revised scheme to establish expected result and performance criteria was simplified to three steps: first GPA to calculate expected significant dimensions, then expected significant pairs of samples in each sensory dimension, and finally expected agreement between assessors and expected agreement with consensus. The difficulty of arriving at proficiency standards cannot be understated. The sensory community has not been able to establish meaningful protocols to ensure panel validity. There is a repetition of the questions, "When is a sensory panel sufficiently trained?" and "At what point is a panelist ready to contribute effectively to a working panel?" as well as "What is the best performance I can expect from a panel working on a specific product category?"

2.4.3. The INRA Approach

The approach taken by Pascal Schlich and his colleagues at INRA (2004) was based on the development of specific SAS macros to perform a range of sophisticated tests that provide a comprehensive analysis of sensory data and deliver it in a form that should be easy for sensory analysts to understand. The subject of panel and panelist performance is fundamental to the usefulness of this sort of sensory analysis. The scheme that is applied to both individual assessors and to panels, as a whole, is based on three performance measures. The first is the ability to repeat results, both as individuals and as a panel. This measure also gets called replication or reliability, but in the instrumental analysis world, it would be called precision. To gain a measure of precision, it is necessary to have replicates. Greater numbers of replicates improve the measurements of variation. In most real-world cases, we are fortunate to have any replicates; however, many sensory labs build two or three replications into the design.

The second measure is accuracy. In the simplest terms, accuracy is the ability to get the correct answer for the value of a specific attribute. To achieve this, the value itself must be known. In the case of the Feedback Calibration method, these values are established in training. However, in most cases, agreement with the panel mean is used as a measure of accuracy. This presupposes that the panel mean is correct. It is also dependent on the attribute in question and the context in which it is being evaluated. The only way to ensure a level of accuracy that permits the comparison of actual results from panel to panel is through calibration. A panel's accuracy is sometimes measured by the homogeneity of its results, or when there are not too many individuals who disagree. If a panel has been trained to specific calibration training targets, it is possible to measure accuracy as distance from a target, distance from a range, adjusted distance from a target, or adjusted distance from a range. In the simplest terms, accuracy can be judged by the number of hits or misses that the panel achieves during a training session. This provides an objective measure that can be compared across panels. However, it must always be put into the context of the degree of difficulty of the attribute within the matrix in which it is evaluated.

This brings us to the third measure, discrimination. Does the individual or panel effectively discriminate different products? Because most of the concern in this area relates to products, not panelists or panels, it will not be elaborated on here. However, with regard to the importance of panel and panelist performance measures, to paraphrase Pascal Schlich, if there's no individual reliability, there can be no panel validity, and consequently no product discrimination.

2.5. Conclusions

The multivariate nature of sensory experience makes multivariate analysis the logical choice to understand the complex relationships that are created by context and the sensory capacity of individual assessors. The use of both simple visualization techniques and more elaborate statistical methods helps to build an accurate and effective picture of the performance of individuals and entire sensory panels. There are different degrees of complexity that can be applied to descriptive analysis data. During the training of panelists, it is best to use rapid methods that help in ensuring that panelists are achieving differentiation of products and mastery of the identification of attributes. Application of rigorous training approaches, such as the Feedback Calibration method, enhances the potential proficiency of both panelists and panels. GPA can give attribute vectors for panelists in a form that is easy to communicate, and it can be performed rapidly. A panel leader may evaluate the progress of his or her panel using the established MANOVA techniques. Panels may be compared and audited using the scheme for panel proficiency or advanced tools such as PanelCheck or the statistical methods of Pascal Schlich.

The objective of all of this effort is to ensure that the data being produced by sensory panel are accurate, precise, and reproducible. Without fulfilling these elements, descriptive analysis cannot profess to be a useful analytical tool.

References

Byrne, D.V., M.G. O'Sullivan, G.B. Dijksterhuis, W.L.P. Bredie, and M. Martens. 2001. Sensory panel consistency during development of a vocabulary for warmed-over flavour. Food Q. Pref. 12:171–187.

Findlay C.J., J.C. Castura, and I. Lesschaeve. 2007. Feedback calibration: A training method for descriptive panels. Food Q. Pref. 18:321–328.

Findlay C.J., J.C. Castura, P. Schlich, and I. Lesschaeve. 2006. Use of feedback calibration to reduce the training time for wine panels. Food Q. Pref. 17:266–276.

Kazi-Aoual, F., S. Hitier, R. Sabatier, and J. Lebreton. 1995. Refined approximations to permutation tests for multivariate inference. Computational Stat Data Analy. 20:643–656.

Lawless, H.T. and H. Heymann. 1998. Sensory evaluation of food, principles and practices. Gaithersburg, MD: Aspen.

McEwan, J.A., E.A. Hunter, L.J. van Gemert, and P. Lea. 2002. Proficiency testing for sensory profile panels: measuring panel performance. Food Q. Pref. 13:181–190.

Naes, T. 1990. Handling individual differences between assessors in sensory profiling. Food Q. Pref. 2:187–199.

Naes, T. 1998. Detecting individual differences among assessors and differences among replicates in sensory profiling. Food Qual. Pref. 9:107–110.

Nilsen, A., O. Tomic, M. Martens, H. Martens, and T. Naes. 2004. The Panel Check—a graphical tool for performance evaluation of sensory panels. Seventh Sensometrics Meeting, Davis, CA.

Pineau, N., R. Pessina, S. Cordelle, A. Imbert, M. Rogeaux, and P. Schlic. 2004. Temporal dominance of sensations: a comparison with conventional profiling and time-intensity. Seventh Sensometrics Meeting, Davis, CA.

Rossi, F. 2001. Assessing sensory panelist performance using repeatability and reproducibility measures. Food Q. Pref. 12:467–479.

Schlich, P. 1996. Defining and validating assessor compromises about product distances and attributes correlations. In T. Naes and E. Risvik (eds.), Multivariate analysis of data in sensory science (pp. 259–306). New York: Elsevier Science.

Schlich, P. 1998. What are the sensory differences among coffees? Multi-panel analysis of variance and FLASH analysis. Food Q. Pref. 9:103–106.

Schlich, P., N. Pineau, D. Brajon, and E.M. Qannari. 2004. Multivariate control of assessors' performances. Seventh Sensometrics Meeting, Davis, CA.

Senstools Version 3.3.1. 2005. OP&P Product Research BV. Utrecht. The Netherlands.

3 A Nontechnical Description of Preference Mapping

3.1. Introduction

Preference mapping (Carroll, 1972) is referred to as a group of multivariate statistical techniques designed to develop a deeper understanding of consumer acceptance of goods. Results of such analyzes can be, for example, used to assist the product developer in selecting from a set of prototypes the single product that may maximize consumer liking. One may wonder why this could not be done by simply looking at the hedonic means for the products tested and selecting the highest-scoring product. However, this practice would overlook the importance of consumer response heterogeneity. It is well known that as consumers, we all have expectations for products that vary, which results in consumers liking different types of products within a category. Ignoring this truth is ignoring the fact that segments of consumers exist, and it is not a recommended practice.

Preference-mapping techniques allow the representation and preservation of the individuality of consumer responses and allow the identification of consumer segments that tend to like the same types of products or have similar expectations for the sensory characteristics of a product. These methods can also provide valuable information about brand positioning and segmentation patterns for the product category being studied (Jaeger et al., 2000). These methods are known as mapping because maps that are multidimensional representations of products based on consumer hedonic responses or sensory characteristics of the products are used. There are two classes of these methods; namely, internal and external preference mapping (EPM). Internal mapping relies on consumer hedonic scores—often overall liking—to determine the multidimensional representation of products and consumers in a common space. External mapping makes use of the sensory characteristics or of an instrumental characterization of the products to provide a multidimensional representation of the products, sensory/instrumental characteristics, and consumers in a common space. In this chapter, we describe in detail both internal and external preference mapping techniques and discuss some of the limitations of the methods and enhancements proposed over the past decade. Chapter 4 builds on techniques described here and concentrates on enhancements to the existing techniques that allow the determination of an ideal sensory profile. Chapter 5 describes some of the same techniques, but in a probabilistic framework.

3.2. Internal Preference Mapping

There have been very good descriptions in the literature of the original algorithm (MDPREF) proposed by Chang and Carroll (1968). Among these is that of Greenhoff and MacFie (1999), which gives a comprehensive review of both internal and external mapping.

As stated earlier, internal preference mapping derives a multidimensional representation of products and consumers. This representation is obtained through singular value decomposition (i.e., principal component analysis) of a data matrix, with products as rows and consumers as variables or columns. For a given product and consumer, the data used are hedonic scores for overall or attribute liking. The principal components (PCs) evaluated are usually referred to as preference dimensions (Greenhoff and MacFie, 1999). Data pretreatment includes centering on the mean for each consumer and often scaling of individuals to unit variance, as suggested by Greenhoff and MacFie (1999). However, some prefer not to normalize each consumer because those consumers who discriminate among products and those who do not are, in the case of normalized data, treated equally. MDPREF can be performed using several software programs, including SAS, SensGear, Senstools, Systat, and MDSX. In this chapter, we use SensGear, as it is a program that we have developed and with which we are most familiar. The data used for this analysis are those in which overall liking data from the tortilla chip example (see Chapter 1) were compiled with products in rows and consumers in columns (i.e., used as variables). Data pretreatment for this analysis consisted of mean centering for columns (i.e., consumers). This is equivalent to performing PCA on the covariance matrix (Borgognone et al., 2001). Greenhoff and MacFie (1999) indicate that columns should also be scaled to unit variance. This practice results in treating all consumers equal with respect to their discrimination of the products and is equivalent to performing the PC analysis on the correlation matrix. This is the same as saying that all consumers would equally influence the determination of the preference dimensions. From a theoretical standpoint, it seems that consumers with small variance do not really express a preference for any of the products and maybe should not be given the same weight as a consumer truly expressing a like for some products and a dislike for others. This can be achieved by performing the analysis on the covariance matrix. Whether or not this makes a significant difference and changes the interpretation of the results is debatable. To address this possibility, we conduct the analysis on both covariance and correlation matrices and show the product scores on the first two preference dimensions (Figure 3.1).

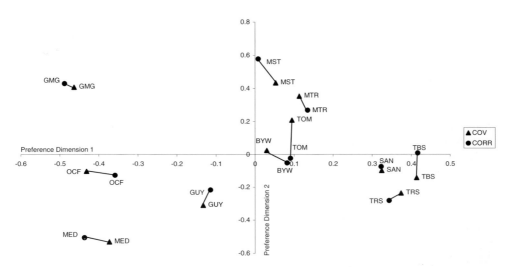

Figure 3.1. A comparison of object (product) scores on preference dimensions 1 and 2 for MDPREF conducted on covariance (▲) and correlation (●) matrices.

Overall, Figure 3.1 shows that the configurations of products on the first two preference dimensions were similar and not greatly affected by data standardization. However, other data sets could show more differences in configuration, especially when the variance for consumers varied greatly or if there were more disagreement from consumers about the most-liked products. Figure 3.2 is a representation of consumers in the preference space for preference dimensions 1 and 2 for a PCA performed on centered overall liking data. The direction of each vector represents the direction of increasing liking for each individual consumer. It is considered to be an approximation because only two dimensions are being considered. The length of the vector is directly proportional to the amount of variance explained by the first two preference dimensions for each consumer. The amount of variance explained for each consumer is assessed by fitting a regression model using the consumer scores as a response variable and the objects' (products) scores on the first two dimensions. On the map, consumers for which the regression model was significant ($\alpha = 0.05$) are plotted.

Details of the singular value decomposition of the centered liking data are given in Table 3.1. Eigenvalues and corresponding percentages of explained variance are given in the table. The data show that the first two preference dimensions explained 42.62% of the variance, and the inclusion of a third preference dimension resulted in 55.33% of the data variance being explained. Corresponding product scores on the PCs (i.e., coordinates used to obtain Figures 3.1 and 3.2) are also given in Table 3.2.

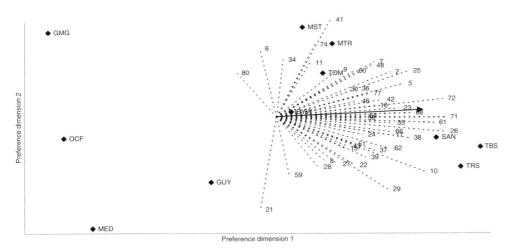

Figure 3.2. Internal preference map (MDPREF) obtained from overall acceptance scores (9-point hedonic scale) given by 80 consumers to 11 white corn tortilla chips. Map obtained from centered data (not normalized).

Table 3.1. Eigenvalues of consumer liking data.

Item	PC1	PC2	PC3	PC4	PC5	PC6	PC7	PC8	PC9	PC10
Eigenvalue	2.91	1.78	1.4	1.2	1.08	0.73	0.63	0.53	0.44	0.3
Percentage	26.44	16.18	12.71	10.94	9.81	6.64	5.7	4.85	3.99	2.74
Cumulative percentage	26.44	42.62	55.33	66.28	76.09	82.73	88.42	93.27	97.26	100

The details of consumer fits are also given in Table 3.3. Of 79 consumers, 47 were significantly fitted, which represents approximately 60% of the original respondents. It is clear from the map that the vast majority of significantly fitted consumers expressed a liking for products SAN (Santita's), TBS (Tostito's bite size) and TRS (Tostito's restaurant style), and many fewer consumers seem to prefer MST (Mission strips) and MTR (Mission triangle). Although one of the main goals of MDPREF is to visually show consumer segmentation, it is quite clear that in the case of white corn tortilla chips, there is little consumer segmentation.

Table 3.2. Principal component (PC) scores.

Product	PC1	PC2	PC3	PC4	PC5	PC6	PC7	PC8	PC9	PC10
BYW	0.78	0.48	0.6	−2.87	−0.02	7.92	−3.68	−4.67	−4.8	1
GMG	−12.07	8.28	3.4	4.18	5.13	4.2	0.01	4.99	1.36	0.47
GUY	−3.46	−6.33	−5.17	−1.45	3.84	2.61	8.49	−2.39	1.16	−1.94
MED	−9.71	−10.84	9.42	−4.19	−4.6	−2.1	−0.26	1.28	0.16	1.37
MST	1.37	8.82	−1.09	−11.56	−0.06	−3.91	0.42	1.96	−1.75	−2.46
MTR	2.95	7.19	7.39	5.95	0.03	−4.98	3.19	−5.5	−0.8	0.88
OCF	−11.24	−2.12	−9.57	3.74	0.91	−5.23	−5.06	−2.07	−0.55	−0.5
SAN	8.44	−2.01	3.46	0.69	0.84	1.59	−4.53	−1.31	5.87	−4.16
TBS	10.76	−2.91	−2.4	−2.41	6.92	−1.74	−0.85	1.56	1.68	5.42
TOM	2.45	4.25	−5.75	1.3	−11.75	2.26	1.6	1	2.83	2.14
TRS	9.72	−4.82	−0.3	6.62	−1.23	−0.6	0.67	5.15	−5.16	−2.23

Table 3.3. Summary of individual consumer fit in preference dimensions 1 and 2.

Consumer	N	R^2	B_0	B_1	B_2	Std B_1	Std B_2	Axis$_1$	Axis$_2$	Significance $p = 0.05$
1	11	0.23	5.55	0.12	0.1	0.99	0.63	2.36	1.51	
2	11	0.04	5.73	0.05	−0.01	0.42	−0.07	0.51	−0.08	
3	11	0.54	6.91	0.16	0.03	1.28	0.17	6.44	0.85	*
4	11	0.47	6.82	0.01	0.17	0.07	1.09	0.36	5.58	*
5	11	0.36	6.27	0.02	−0.15	0.17	−0.99	0.74	−4.2	
6	11	0.38	5.36	0.2	−0.01	1.68	−0.05	4.54	−0.13	*
7	11	0.16	6	0.05	0.07	0.41	0.46	1.28	1.44	
8	11	0.45	6	0.09	0.05	0.72	0.33	4.86	2.25	*
9	11	0.4	7.82	−0.03	0.09	−0.27	0.55	−2.12	4.28	*
10	11	0.62	7	0.05	0.05	0.4	0.29	6	4.38	*
11	11	0.21	7.09	0.06	0.07	0.52	0.47	1.89	1.7	
12	11	0.12	6	0.09	−0.05	0.7	−0.35	1.27	−0.64	
13	11	0.62	5.73	0.13	0.08	1.1	0.53	6.69	3.22	*
14	11	0.56	5.73	−0.02	0.18	−0.15	1.16	−0.88	6.69	*
15	11	0.62	5.82	0.17	0.23	1.4	1.46	5.17	5.4	*
16	11	0.42	6.82	0.05	−0.1	0.38	−0.65	2.54	−4.31	*
17	11	0.47	6.73	0.09	0.16	0.71	1	3.25	4.6	*
18	11	0.79	6.55	0.16	−0.14	1.29	−0.87	7.82	−5.3	*
19	11	0.46	6.82	0.04	0.14	0.3	0.91	1.78	5.28	*
20	11	0.2	6.36	0.03	0.11	0.27	0.7	0.85	2.23	
21	11	0.27	6.55	0.09	0.06	0.71	0.35	2.94	1.48	
22	11	0.4	6	0.08	−0.09	0.64	−0.55	3.61	−3.08	*
23	11	0.31	6.36	0.13	0.09	1.1	0.6	3.29	1.8	
24	11	0.52	7.09	0.08	−0.03	0.65	−0.2	6.01	−1.8	*
25	11	0.44	6.27	0.16	0.04	1.31	0.27	5.2	1.09	*

(Continued)

Table 3.3. *Continued.*

Consumer	N	R^2	B_0	B_1	B_2	Std B_1	Std B_2	Axis$_1$	Axis$_2$	Significance $p = 0.05$
26	11	0.19	7	0.07	−0.03	0.59	−0.2	2.22	−0.73	
27	11	0.13	8.27	−0.03	−0.04	−0.21	−0.24	−0.99	−1.13	
28	11	0.75	5.36	−0.03	−0.41	−0.25	−2.63	−0.87	−9	*
29	11	0.52	7.55	0.05	−0.08	0.44	−0.51	4.12	−4.69	*
30	11	0.41	5.91	0.14	−0.07	1.14	−0.44	4.56	−1.76	*
31	11	0.69	5.36	0.18	0.15	1.47	0.95	6.94	4.47	*
32	11	0.75	6.27	0.27	−0.06	2.23	−0.37	8.91	−1.46	*
33	11	0.47	5.82	0.11	−0.2	0.89	−1.27	3.21	−4.59	*
34	11	0.45	4.82	0.08	−0.22	0.64	−1.41	2.21	−4.86	*
35	11	0.76	6.73	0.15	−0.22	1.2	−1.44	5.86	−7.03	*
36	11	0.37	7.18	0.07	0.07	0.58	0.42	3.64	2.64	*
37	11	0.05	4.64	0.05	0.01	0.4	0.07	0.63	0.11	
38	11	0.33	5.18	0.05	0.15	0.39	0.98	1.44	3.67	
39	11	0.16	7.09	0	−0.03	−0.02	−0.22	−0.2	−1.92	
40	11	0.51	6.18	0.11	−0.02	0.94	−0.1	6.09	−0.62	*
41	11	0.42	7.36	0.09	0.08	0.78	0.5	4.23	2.72	*
42	11	0.51	5.64	0.21	−0.17	1.69	−1.07	5.17	−3.28	*
43	11	0.61	5.82	0.27	−0.1	2.26	−0.67	6.97	−2.07	*
44	11	0.51	6.55	0.08	−0.08	0.62	−0.52	4.71	−3.93	*
45	11	0.14	6.91	0.1	0.02	0.81	0.1	1.7	0.22	
46	11	0.83	5	0.02	0.11	0.2	0.68	2.86	9.58	*
47	11	0.48	6.64	0.21	0.08	1.69	0.51	5.56	1.66	*
48	11	0.4	6.73	0.11	−0.11	0.92	−0.71	3.76	−2.89	*
49	11	0.12	5.82	0.05	0.02	0.38	0.13	1.32	0.47	
50	11	0.37	6.36	0.15	0.07	1.27	0.45	4.24	1.49	*
51	11	0.26	5.64	0.11	−0.09	0.92	−0.57	2.67	−1.67	
52	11	0.59	5.64	0.12	0.16	1.01	1.01	5	5.01	*
53	11	0.14	6.36	0.03	0.08	0.27	0.5	0.8	1.48	
54	11	0.01	5.55	0.02	−0.02	0.15	−0.11	0.13	−0.1	
55	11	0.4	7.09	0.09	−0.07	0.73	−0.48	4.04	−2.65	*
56	11	0.24	6.36	−0.04	0.06	−0.35	0.37	−2.04	2.11	
57	11	0.04	6.09	−0.01	−0.03	−0.1	−0.19	−0.22	−0.46	
58	11	0.04	7.18	−0.01	−0.03	−0.06	−0.2	−0.14	−0.42	
59	11	0.12	6.27	−0.06	0.07	−0.52	0.45	−1.08	0.94	
60	11	0.16	7.64	−0.05	0.01	−0.37	0.04	−1.96	0.23	
61	11	0.34	7.36	0.05	−0.04	0.4	−0.26	3.43	−2.21	
62	11	0.47	6.73	0.01	−0.1	0.07	−0.62	0.67	−5.66	*
63	11	0.5	6.64	0.09	0.13	0.75	0.82	4.06	4.47	*
64	11	0.69	5.82	0.25	−0.02	2.1	−0.15	8.31	−0.58	*
65	11	0.56	6.91	0.13	−0.09	1.09	−0.56	5.95	−3.07	*
66	11	0.3	4.82	0.09	−0.16	0.75	−1.03	2.16	−2.94	
67	11	0.39	6.64	0.09	0	0.75	0.01	4.64	0.09	*
68	11	0.23	5.64	0	0.18	0.01	1.13	0.04	2.82	
69	11	0.51	4.73	0.2	−0.06	1.62	−0.4	6	−1.47	*
70	11	0.59	7.36	0.09	0.01	0.71	0.04	7.08	0.39	*
71	11	0.03	7.45	−0.02	−0.01	−0.16	−0.07	−0.36	−0.17	
72	11	0.23	6.73	0	0.05	−0.02	0.31	−0.2	2.71	
73	11	0.74	7.82	0.06	0	0.52	0	8.9	−0.04	*
74	11	0.75	6	0.26	0.07	2.18	0.44	8.78	1.75	*
75	11	0.18	5.45	0.12	0	1.02	0.01	2.11	0.03	
76	11	0.62	5.27	0.07	0.31	0.56	1.97	2.02	7.1	*
77	11	0.26	5.82	0.11	0.06	0.91	0.36	2.84	1.14	
78	11	0.31	6.55	0.11	0.06	0.92	0.4	3.39	1.46	
80	11	0.15	5	0.07	0.01	0.6	0.05	1.83	0.14	

To illustrate the concept of consumer segmentation further using MDPREF, the same analysis was performed on overall liking for 10 muscadine grape juices (see section 1.2 for details of the study) by 61 consumers (Figure 3.3). The analysis clearly demonstrates that consumers showed preferences for different products. This is because consumer vectors are not concentrated in one direction of the map but are instead distributed in all four quadrants. Vectors pointing toward the upper-right quadrant indicate that consumers liked products Carlos and Black Beauty most, whereas consumers in the lower-right quadrant preferred Black Beauty, Granny Val and Summit. Although the first two preference dimensions explained only about 48% of the variation in consumer hedonic scores, 74% (45 of 61) of the consumers were significantly fitted ($\alpha = 0.05$) in the space. The overall preference direction considering this consumer group as a single segment was toward the bottom quadrant of the map. A simple segmentation of consumers according to the location of their preference vectors in a particular quadrant of the map shows that these consumers expressed liking for different products (Figure 3.4). Segment 1 (consumers whose preference vector pointed toward the upper-right side of the map) liked Black Beauty, Carlos, and Post White best, whereas segment 2 liked Granny Val best, segment 3 liked Post Red, and segment 4 liked Ison, Post Red, Southern Home, and Supreme most. This could have also been intuitively deducted by projecting each product on a vector going through the middle of each of the four quadrants. We will not illustrate here the concept of consumer segmentation further, as this will be a subject later addressed in this book (see Chapter 6).

McEwan (1998) described some improvements to the existing method of MDPREF. She described extended internal preference mapping, which features a clustering of consumers pre-MDPREF and a projection of the external data such as sensory descriptive data into the preference space. The projection of external data into the preference space is most interesting because it helps with the interpretation of the preference dimensions obtained with MDPREF. This projection is performed by regressing the attribute mean scores onto the product coordinates or scores in the preference space (Greenhoff and MacFie, 1999). An example of such projections is given in Figure 3.5 for the muscadine

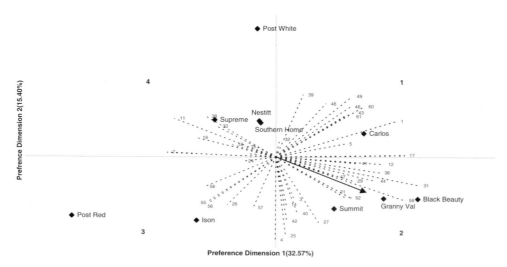

Figure 3.3. Internal preference map (MDPREF) obtained from overall acceptance scores (9-point hedonic scale) given by 61 consumers to 10 muscadine grape juices. Map obtained from centered data.

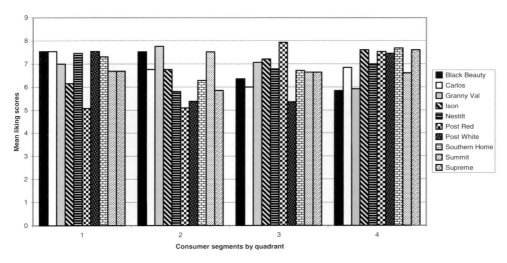

Figure 3.4. Mean overall liking scores for segments of consumers, formed on the basis of the direction of preference vectors (1 = upper-right, 2 = lower-right, 3 = lower-left, and 4 = upper-left quadrants).

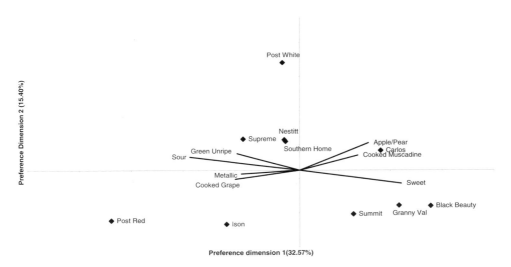

Figure 3.5. Projection of external data (i.e., sensory attributes) in the preference space (MDPREF).

grape juice sample data set. In the figure, only attributes significantly fitted ($\alpha = 0.05$) are displayed. Attributes projecting in similar directions were most highly positively correlated to each other, whereas attributes projecting in opposite directions were most highly negatively correlated. The overall preference direction being toward the lower-right quadrant, sweetness is interpreted to be the attribute most highly correlated with the average acceptance of the products, meaning that sweetness is a driver of product liking. However, the projections of sensory attributes also provide some clues as to differences in liking observed among consumers. The map shows that consumers with preference vectors pointing toward the right side of the map liked sweet juices rich in apple/pear and cooked muscadine notes, whereas consumers on the left side of the map liked products that were more sour and higher in cooked grape and green/unripe notes. This seems to be a very valuable enhancement to the original MDPREF method, especially from the map interpretation standpoint.

One of the problems in internal preference mapping (MDPREF) is that clusters of consumers can be difficult to identify on the basis of the singular value decomposition of centered hedonic data. To overcome this, preference-clustering techniques have been employed (Martinez et al., 2002; McEwan, 1998). A classical application of these techniques is that of first clustering consumers according to their product acceptability patterns (Martinez et al., 2002), using Ward's hierarchical clustering with Euclidian distances on the unstandardized hedonic data. The resulting cluster hedonic averages are then submitted to a PCA analysis, and sensory attributes are projected in the space, as previously described. Results of such an analysis are given in Figure 3.6. Because MDPREF is performed on the average liking scores of the five different clusters selected, the resulting preference space is somewhat different from that given by Figure 3.3. Although most products are in a similar location (e.g., Post White, Black Beauty, Summit, Post Red, and Granny Val), other products such as Summit and Carlos are located in slightly different regions of the map. This makes the interpretation of the map slightly different. In addition, the clusters that could intuitively be identified in Figure 3.3 do not seem to be well represented in the extended map with clusters (Figure 3.6a). The largest cluster (i.e., two) is located in the center of the map, and it may be hard to determine what these consumers liked about the products presented to them. However, cluster 2 is better represented in preference dimensions 1 and 3, where it is evident that cluster 2 was made of consumers liking products such as Summit and Post Red.

Figure 3.7 displays the projected sensory attributes in the preference space. The attribute projections and the significant drivers of preference were essentially the same between the MDPREFs performed on cluster means and the standardized data for individuals. To understand the results of Ward's cluster analysis and their relationship to MDPREF performed on standardized consumer data (i.e., the MDPREF presented in Figure 3.3), consumers' cluster numbers were identified on the original MDPREF map (Figure 3.8). Results indicate that Ward's hierarchical clustering grouped together individuals in cluster 2 whose preference vectors were vastly different (i.e., pointing in various directions) for the first three preference dimensions (Figure 3.8a, 3.8b). This shows that extended preference mapping as defined by McEwan (1998), in which consumers are first clustered and then an MDPREF is performed on the product means for the clusters, is not always consistent with MDPREF performed on raw data (i.e., centered or standardized). The authors believe that clustering extended internal preference mapping could yield inadequate interpretations. We believe this to be because clusters are often of different sizes, which means that small clusters may not be adequately representing reality. Therefore, unless the number of consumers is very large and the clusters are of relatively equal sizes, we do not recommend the use of extended internal preference mapping. Instead, we feel that a visual inspection of the MDPREF performed on raw consumer data is most appropriate. However, additional research on clustering techniques may yield clusters that may be more representative of MDPREF maps.

Courcoux and Chavanne (2001) proposed a latent class approach to MDPREF that has the advantage of concurrently clustering consumers into a small number of segments and representing the clusters in the product space. This approach yields maps that are interpreted similarly to classical MDPREF. The latent class vector model assumes the existence

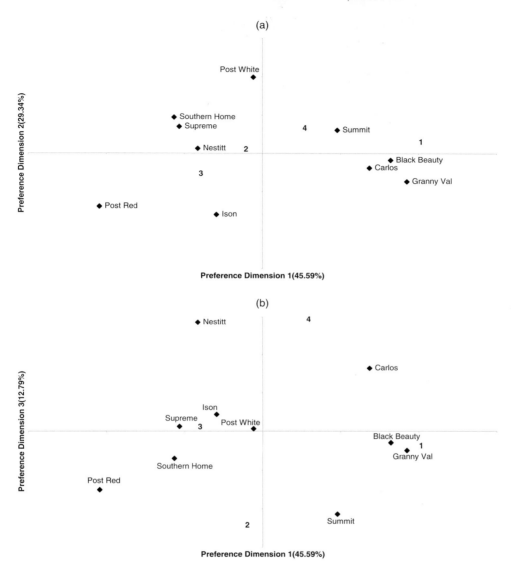

Figure 3.6. Extended internal preference mapping for four clusters of consumers. Preference dimensions created using centered average consumer data (i.e., for each cluster). Cluster analysis performed using Ward's hierarchical clustering with Euclidian distances. Fonts representing clusters are proportional to cluster size. (a) PC1 versus PC2. (b) PC1 versus PC3.

of T classes of consumers. In each class, the preference vector for each consumer for the products is assumed to be a random variable normally distributed. Estimation procedures for the model are based on the maximum likelihood, and a latent class approach is used for the determination of segments using an expectation maximization algorithm, as described by Dempster et al. (1977) and McLachlan and Basford (1998). Details of the estimation procedure are found in De Soete and Winsberg (1993).

Figure 3.7. Projection of external data (i.e., sensory attributes) in the preference space established on clustered data (extended MDPREF). (a) PC1 versus PC2. (b) PC1 versus PC3.

3.3. External Preference Mapping

As stated earlier, external preference mapping (PREFMAP) derives a multidimensional representation of products based on their sensory profile or a set of other external data such as instrumental measures of color, texture, or flavor. This representation is obtained through singular value decomposition (i.e., principal component analysis) of a data matrix, with products as rows and external data as variables or columns. External mapping approaches are limited by the fact that the sensory space (i.e., multidimensional representation) is obtained from external data alone, without prioritization of the attributes based on their importance to consumers. This implies that the representation obtained may have no relevance to consumer acceptance of the products. This is, in our opinion,

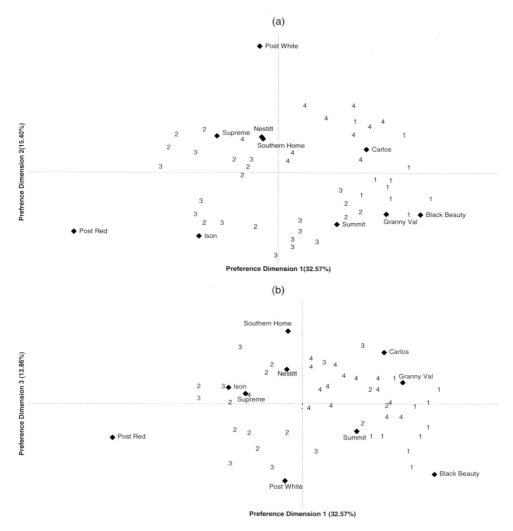

Figure 3.8. Internal preference mapping performed on centered consumer data. Cluster analysis performed post-MDPREF, using Ward's hierarchical clustering with Euclidian distances. Each consumer preference vector is represented on the map by the customer's cluster number.

a major limitation because not all attributes that vary in intensity in a set of stimuli will have a significant effect on acceptance by consumers. With that said, external preference mapping is a popular methodology—maybe even more popular than MDPREF. The first step in EPM is to provide a multidimensional representation of the sensory characteristics of the products, usually though PC analysis. The data matrix is made of products as rows and sensory attributes as columns. For each product and each attribute, the mean sensory panel score is the input. This implies that panel consonance (i.e., variance from the mean) is not taken into account in this type of analysis. However, this will later be addressed in a different context in Chapter 8, which deals with the application of risk analysis methodologies to the prediction of consumer preference. As for internal preference mapping, we first face the decision of whether to use centered or standardized data. Borgognone et al. (2001) provide an excellent discussion on the subject of using centered (covariance matrix) versus standardized (correlation matrix) data. Their paper included a survey of the literature that showed that a majority of authors used standardized data to perform the PC analysis of sensory data. However, the authors argue against this practice, and we tend to concur with their argument. The use of the covariance matrix (centered data) implies that only the variation in the mean across attributes is removed from the data, but the variance across products for the different attributes is not considered equal (i.e., this is the case when the data are standardized). It seems that the idea of variance is an important one in sensory data and should not necessarily be removed from the analysis. It is recommended that centered data be used if the sensory attributes are expressed in the same units. However, if attributes are measured on different scales or the magnitude of the scores across attributes varies greatly, standardized data may be more appropriate. In actuality, what happens in a PCA on centered data is that attributes with larger variance will be favored over attributes with small variance.

To illustrate the difference in the maps created with centered and standardized data, a case study on the sensory properties of white corn tortilla chips (see section 1.1) is presented. Figure 3.9 is a comparison of variable (i.e., sensory attribute) loadings on the first two PCs for centered (Figure 3.9a) and standardized (Figure 3.9b) sensory means. For the space derived from centered data, the first and second PCs explained 58.6% and 19.0% of the variability in the original variables, respectively. The attributes loading most highly on PC1 were salt, toasted corn, and scorched with positive loadings, and toasted grain and masa with negative loadings. For PC2, scorched, toasted grain, and cardboard loaded positively, whereas salt and masa loaded negatively. All other attribute loadings were very small, indicating that these attributes played only a small role in dictating product locations in the map. For standardized data (Figure 3.9b), the attribute loadings were quite different, with most attributes loading highly on at least one of the first two PCs. This implies that many more of the original attributes will dictate the position of the products in the sensory space.

A comparison of the product locations in the two derived sensory spaces is given in Figure 3.10. Results clearly show that the spatial representation of the samples on the first two PCs is quite different when using either the correlation or the covariance matrices. This is especially important because the coordinates of the products in the space (product scores) will be used to fit individual consumers. Because PCA solutions are not unique and can be rotated, this disagreement could be in part reduced by rotations or using some sort of scaling. To investigate this possibility, the product scores on the various PCs were subjected to generalized procrustes analysis. Graphic results of this analysis are presented in Figure 3.10a for when the first two PCs were used in the analysis and in Figure 3.10b when the product scores on the first five PCs were used. Figure 3.11b demonstrates that the product configurations for the first two PCs obtained using either the correlation or the covariance matrices were very different from each other, even when accounting for possible rotation and scaling. However when the product configurations compared included more PCs (five in this case), the two product configurations were much more similar.

However, as the first two or three PCs are usually used in external preference mapping, the data matrix used to generate the PCA greatly matters. Although this chapter does not intend to establish which solution is more sensible, it would seem to be an important subject for future research.

The second step in the analysis is to fit the consumer data in the sensory space. To do this, some type of polynomial model is used to regress the hedonic scores given to the

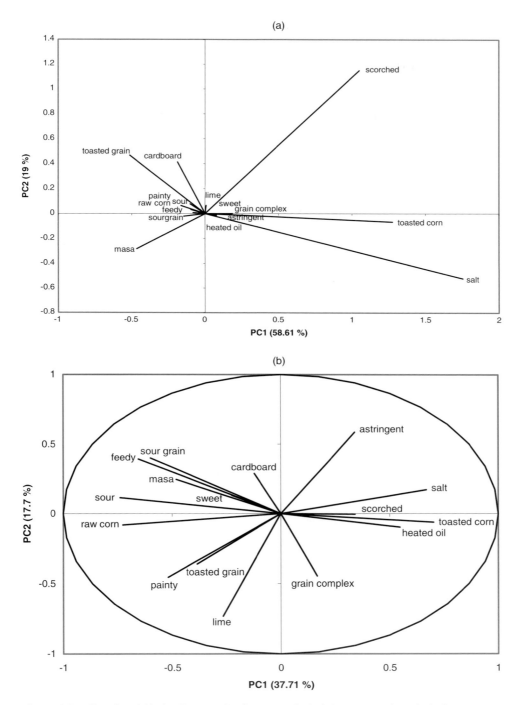

Figure 3.9. Plot of variable loadings on the first two principal components for principal component analysis performed on a (a) covariance matrix and (b) correlation matrix.

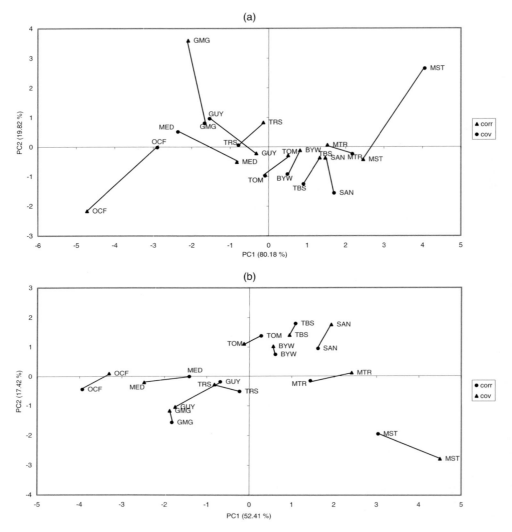

Figure 3.10. Generalized procrustes analysis performed on the product scores on principal component analysis performed on the covariance and correlation matrices. (a) Product scores on the first two principal components used. (b) Product scores on the first five principal components used.

products onto the coordinates of the products in the sensory space. Four possible models (i.e., vector, circular, elliptical, and quadratic) were originally described by Carroll (1972). The models considered are special cases of the quadratic surface model which can be expressed as:

$$OL = \sum_i a_i PC_i + \sum_i b_i PC_i^2 + \sum_{ij} c_{ij} PC_i PC_j, \qquad \text{(Eq. 3.1)}$$

where OL is overall liking (hedonic scores), PC_i are the product scores on the ith PC, a_i and b_i are regression coefficients for the ith PC linear and quadratic terms, and c_{ij} is the regression coefficients for the interaction between the ith and jth PCs. For the simplest

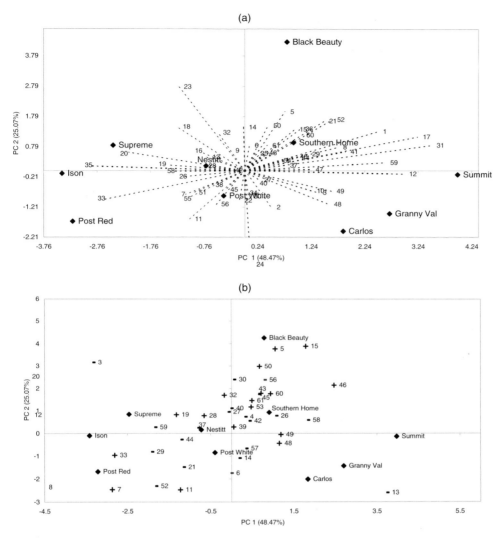

Figure 3.11. External preference mapping performed on the standardized sensory data of muscadine juices. Consumers fitted with (a) vector model and (b) circular model. Only those consumers who were significantly fitted (p = 0.1) are represented in the figure.

model, the vector model, b_i and c_{ij} are equal to zero. For the circular model, b_i is constant across PCs, and c_{ij} is zero. For the elliptical model, only c_{ij} is zero, and no constraints are placed on b_i.

In the original PREFMAP method, the more complex model (i.e., quadratic) was first fitted to each consumer. Models of decreasing complexities were then fitted and compared for fit to the quadratic model. The simplest model, which was not of significantly lower fit than the quadratic model, was retained as the final model for a particular individual. One of the weaknesses of external preference mapping is that only a small proportion of consumers are sometimes significantly fitted by any of the models. For example, Jaeger et al. (2000) gave an example in which only 30% of consumers were significantly fitted. Jaeger et al. (2000) also showed a slightly greater number of consumers (30% vs. 27%) who were significantly fitted to the sensory space when the correlation matrix was used to shape the sensory space. A similar approach was taken here to determine the effect of sensory data standardizing on the percentage of consumers significantly fitted in the space. The analysis was performed with SensGear, using first only the vector model to fit consumers in the sensory space generated on either centered or standardized sensory data. In a second set of analyses, the auto–best fit option was used to allow fitting of the more complex models when necessary.

Results (Table 3.4) indicate that, for the tortilla chip data, a greater percentage of consumers (41%) were fitted by the vector model when the correlation matrix of the sensory data was used to perform the PCA than when the covariance matrix was used (21%). However, the same was not true for the muscadine juice data (31% vs. 30%). The use of the auto–best fit option in SensGear, which starts by fitting the quadratic surface model and finds the simplest model not significantly affecting the quality of the fit, resulted in an increase in the proportion of consumers significantly fitted (i.e., from 3 to 12 percentage points, depending on the type of matrix used and the data set). These results point out one of the major limitations of external preference, which is that a low proportion of consumers are usually fitted adequately (i.e., significantly) in the space. In the two examples given here, only 43% and 44% of consumers were significantly fitted on the first two sensory dimensions for the tortilla chip and muscadine juice data sets, respectively.

Improvements to external preference mapping have been proposed by Faber et al. (2003). One of the common criticisms of EPM has been that only the first two PCs are used in fitting consumers to the sensory space. This is because the addition of more PCs with circular, elliptical, or quadratic models would result in overfitting the data (i.e., one would run out of degrees of freedom). The use of the first two PCs results in the poor fitting of a considerable number of consumers. This could lead to decisions being made about product optimization based on a small proportion of consumers and does not necessarily inspire confidence in EPM. Faber et al. (2003) proposed to restrict EPM to the use of linear models (i.e., vector models) but to include a greater number of PCs in fitting consumers to the sensory space. In an example of EPM for apples, the authors showed an improvement in consumers adequately fitted to the space from 51% for two PCs to 80% for five PCs. This simple improvement of the original methodology may help the user feel more confident about the clusters of consumers found. This principle applied to our two data sets using only the first three PCs resulted (Table 3.4) in an improvement of the

Table 3.4. Percentage of significantly fitted consumers for external preference mapping, using either covariance and correlation matrices and vector or more complex models.

		Data	
Matrix[1]	**Model**[2]	**Tortilla chips**	**Muscadine juices**
Correlation	Vector	41[3]	31
Covariance	Vector	21	30
Correlation	Auto	44	38
Covariance	Auto	26	43
Correlation	Vector (3PC)	57	38
Covariance	Vector (3PC)	57	61

[1]Type of matrix used to generate the sensory space.
[2]Type of model fitted to consumers. Vector = vector model with two PCs; auto = best of quadratic, elliptical, circular, and vector; vector (3PC) = vector model with three PCs.
[3]Percentage of consumers significantly fitted ($\alpha = 0.1$).

proportion of consumers adequately fitted to levels of 57% and 61% for the tortilla chip and muscadine juice preference maps, respectively. Examples of graphic outputs of external preference maps are given in Figure 3.11a for the vector model and Figure 3.11b for the circular model. The circular model was selected here so that ideal points for consumers could be identified. An important characteristic of the circular model, as well as of the elliptical and quadratic surface models, is that the minimum or maximum predicted value for a particular individual can be determined. The concept of an ideal point is quite simple. It assumes that in the space, there exists a location that would maximize or minimize liking by a particular consumer. This implies that there is a combination of sensory attribute levels that would represent the ideal product. In Figure 3.11b, a negative sign represents a consumer negative ideal or point at which liking is minimal for that consumer, whereas a positive sign represents the location on the map that maximized liking for a particular consumer. There were a total of 51 ideals represented on the map, with 23 being positive and 28 being negative. The vector model representation in Figure 3.11a shows that 49 of 61 consumers were significantly ($p < 0.1$) fitted in the space by the vector model. Each vector represented indicates the general direction of increasing liking for each consumer. However, the concept of ideal point does not apply for the vector model.

External preference mapping is impaired by reliance on sensory profiles given by a trained panel to produce an appropriate multidimensional representation (i.e., from a consumer point of view) of the stimuli (i.e., products). Jaeger et al. (2000) point out that for external analysis to be successful, it is essential that the external stimulus space contain dimensions that pertain to preference. The authors argue that preference mapping can be improved by taking into account the behavioral process taking place in the formulation of a hedonic judgment. Their point of view is based on the fact that some of the sensory information describing similarity and differences between the products is not used uniformly in the synthesis process of consumers. In other words, some sensory attributes drive liking more than others. The authors propose to weight the sensory attributes before a multidimensional representation (i.e., PCA) of the profiles is determined. This was performed by first employing extended internal preference mapping, as described by McEwan (1998). In the extended version of internal mapping, the sensory attributes are projected on the space derived from hedonic data. For each attribute, a correlation between the sensory attribute scores of the products and the projected distances between products on the sensory vector can be assessed. It seems logical that the sensory attributes best fitted in this space would be those driving liking and should be given a greater weight (synthesis weight) in the establishment of the sensory space derived from the sensory profiles. The application of this method to a preference map of apples provided some significant improvements in the percentage of consumers adequately fitted in the sensory space by a vector model when the sensory attributes were weighted according to their correlation to the preference dimensions obtained by MDPREF. To evaluate this method improvement, both the white corn tortilla chip and muscadine juice data sets were submitted to such an analysis (Table 3.5). Three different strategies were employed to assign weight to sensory variables before PCA. First, no weights were assigned; this was the control PREFMAP. Second, the weights were assigned as the correlation (absolute value) between the attribute scores for the products and the product scores on the first two preference dimensions of an MDPREF (i.e., attribute projection). Third, weights were assigned as the same correlation as in the second strategy, but only if the attribute projection were significant ($\alpha = 0.05$); otherwise, the attribute weight was set to zero. In addition, results were compared with calculations made on both the correlation and covariance matrices and also compared with the percentage of subjects significantly fitted ($p < 0.1$) in the MDPREF. Results

Table 3.5. Percentage of significantly fitted consumers for external and internal preference mapping, using either covariance or correlation matrices and various weighting strategies of the sensory data for external mapping.

		PREFMAP		
Data and matrix	MDPREF	Weight = 1	Weight = R	Weight = R if significant ($\alpha = 0.05$)
Tortilla Chips				
Correlation	81	57	57	66
Covariance	70	57	58	59
Muscadine Juices				
Correlation	74	38	38	49
Covariance	66	61	56	56

indicate that in all cases, MDPREF allowed for a greater proportion of subjects to be fitted (66%–81%) than with any of the PREFMAP methods investigated. When comparing the percentage of consumers fitted when no weighting of the sensory data was used (i.e., traditional PREFMAP), the two weighting strategies did not provide any improvements when the PREFMAP was performed on the covariance matrix. The third weighting strategy performed on standardized data was the only case for which there was an increase in the number of consumers fitted in the space, from 57% to 66% for the tortilla chip data and from 38% to 49% for the muscadine juice data. For the two examples given here, weighting of the sensory data did not provide a magic solution to some of the inherent fitting issues associated with PREFMAP. However, weighting was found to provide some improvements and to be a step in the right direction.

3.4. Conclusions

We have outlined in this chapter the two types of methodologies available to perform analyses known as preference mapping. We have shown that MDPREF tended to better represent consumer liking patterns in the sensory space than did PREFMAP. The question that remains about both approaches is with regard to actionability in an industrial setting. MDPREF is attractive because it can provide information about consumer segmentation. It seems that this type of information could shape decisions made about the development of new products, especially if the demographic type of data are useful in describing consumer clusters. The projection of external data in the MDPREF space is also useful to determine what attributes drive liking by consumers. For that matter, PREFMAP gives the same type of information except for the fact that consumers can be described by more complex models in the space. Although these methods provide information about the relationship between liking by groups of consumers and sensory attributes, the optimal level of a specific attribute is not necessarily identified. From the standpoint of product development, this causes considerable problems. It is one thing to find through MDPREF or PREFMAP that saltiness drives liking, it is yet another to determine how much salt is enough or how much is too much. This is where we see preference mapping in its original forms fail to provide enough information to the product developer to formulate an optimal product from the sensory standpoint. This is the subject of the next two chapters, as we describe there the methods that are extensions of preference mapping—some internal, some external, some deterministic, some probabilistic—that will allow the identification of the sensory characteristics of an optimal product or products.

References

Borgognone, M.G., J. Bussi, and G. Hough. 2001. Principal component analysis in sensory analysis: covariance or correlation matrix. Food Q. Pref. 12:323–326.

Carroll, J.D. 1972. Individual differences and multidimensional scaling. In R.N. Shepard, A.K. Romney, and S.B. Nerlove (eds.). Multidimensional scaling: theory and applications in the behavioral sciences (pp. 105–155). New York: Seminar Press.

Chang, J.J. and J.D. Carroll. 1968. How to use MDPREF, a computer program for multidimensional analysis of preference data. Unpublished report, Bell Telephone Laboratories.

Courcoux, P. and P.C. Chavanne. 2001. Preference mapping using a latent class vector model. Food Q. Pref. 12(2001):369–372.

Dempster, A.P., N.M. Laird, and D.B. Rubin. 1977. Maximum likelihood estimation for incomplete data via the EM algorithm. J. R. Stat. Soc. Ser. B 39:1–38.

De Soete, G. and S. Winsberg. 1993. A latent class vector model for preference ratings. J. Classification 10:195–218.

Faber, N.M., J. Moyet, and A.A.M. Poelman. 2003. Simple improvements of consumer fit in external preference mapping. Food Q. Pref. 14(2003):455–461.

Greenhoff, K. and H.J.H. MacFie. 1999. Preference mapping in practice. In H.J.H. MacFie and D.M.H. Thomson (eds.) Measurement of food preferences. Gaithersburg, MD: Aspen.

Jaeger, S.R., I.N. Wakeling, and H.J.H. MacFie. 2000. Behavioural extensions to preference mapping: the role of synthesis. Food Q. Pref. 11(2000):349–359.

Martinez, C., M.J. Santa Cruz, G. Hough, and M.J. Vega. 2002. Preference mapping of cracker type biscuits. Food Q. Pref. 13(2002):535–544.

McEwan, J.A. 1998. Cluster analysis and preference mapping (review no. 12, project no. 29742). UK Campden & Chorleywood Food Research Association.

McLachlan, G.J. and K.E. Basford. 1988. Mixture models. New York: Marcel Dekker.

4 Deterministic Extensions to Preference Mapping Techniques

4.1. Introduction

As sensory scientists or product developers, we are often faced with the challenge of creating products with maximum acceptance levels by consumers. In any product optimization process, one is forced to develop an understanding of the relationship between sensory intensity for a specific attribute and the hedonic level or liking for the particular sensory intensity. This relationship, described many times in the literature (Moskowitz, 1985), is often best described by an inverted U-shaped function. This type of relationship implies that there is—for any sensory attribute—an optimal level, below and above which liking for the product is less than that for the optimal sensory level. Although this representation is often simplistic because it ignores interactions (i.e., correlations) among attributes, it can be useful when only one or two sensory dimensions are being optimized. For example, if an optimal salt level for our tortilla chip data were to be determined, one could vary the salt level through an experimental design, keep all other attributes constant, and determine the characteristics of the inverted U-shaped function. However, most will agree that this is not practical in many situations, including when, for example, the number of sensory dimensions is too large, rendering the experimental design–based approach too costly or too time consuming.

There are many methods that have been described over the years that, based on a carefully selected set of products (i.e., often a combination of commercial products and prototypes), allow the identification of an optimal sensory profile that will maximize consumer liking. These methods are usually referred to as preference mapping techniques; they are extensions of the methods originally proposed by Carroll and others in the 1960s (Carroll, 1972). In Chapter 3, we undertook the description of more classical techniques such as MDPREF and PREFMAP; here we describe regression-based techniques that allow the identification of sensory attributes driving liking of products by consumers.

4.2. Application and Models Available

Numerous regression methods exist for relating consumer data to descriptive profiles, including ordinary least squares regression, principal component regression, and partial least squares regression (PLSR). In this type of study, a minimum of six products is tested, and often many more. The experimenter selects products that are commercial products, prototypes, or a combination of the two. The same products are presented to a group of consumers to assess the hedonic level (i.e., liking) of each product. The questions asked of consumers can be about their overall liking for the products or about the liking of more specific attributes such as appearance, flavor, or texture. However, if liking data are gathered on specific attributes, correlations among attribute liking ratings are often very high (i.e., a product liked overall also will score high on appearance, flavor, and texture, whereas a product disliked overall will have a tendency to score low on all specific attributes). This is why in the examples given here we concentrate on modeling overall liking. In addition to consumer data, external data are needed from a descriptive panel or instrumental measures, when instrumental methods to assess the appearance, flavor, and texture attributes of the products are available.

4.2.1. PLSR on Average Data

For the application of PLSR, modeling can occur at the consumer level or the panel level. At the panel level, data are averaged for each product across panelists for the descriptive data and across consumers for the hedonic data, and a data matrix is created, with products as rows and hedonic and predictive (sensory or instrumental) variables as columns. This implies that means are appropriate to be taken, assuming that consumer data were homogeneous (i.e., consumers agreed as to the rank order of the products and panelists agreed about the intensities given to each product). If there is evidence that consumers disagreed about their relative liking for the products, a segmentation technique such as cluster analysis should be employed. Unfortunately, there is not a simple fix to sensory panel disagreement except for the fact that when there is disagreement or lack of repeatability, the analysis of variance may yield a nonsignificant product effect (i.e., no significant differences among products). We suggest not including the attributes for which there is no significant difference among products. In simple terms, if all products have a constant level of one attribute, that attribute could not have affected the differences observed among the products for liking. We would also point out that consumer data does not have to be averaged if the PLS2 algorithm of PLSR is used.

Once the data matrix is established, the data should be in a format similar to that presented in Table 4.1. The principle of regression applied to the data is very simple. What is attempted is to find a function of the predictive variables (sensory or instrumental) that will best describe the variation in the hedonic data.

To exemplify this type of analysis, PLSR was selected here because we feel that PLS is one of the most appropriate multivariate regression methods available. We are not using ordinal least squares regression for the obvious reason that we have more predictors (i.e., sensory variables) than we have observations (i.e., products). We therefore have to rely on data decomposition techniques. In PLSR and the other algorithms available, the original

Table 4.1. Mean consumer acceptance and descriptive profiles.

Product Codes	Overall Liking	Visual Attributes		Flavor Attributes		Texture Attributes	
		Degree of Whiteness	Grain Flecks	Salt	Grain Complex	Hardness	Crispness
BYW	6.4	6	8	8.8	6.9	8.8	10.4
GMG	5.5	2	3	7.2	7.2	9.6	9.6
GUY	5.9	7	6	7.8	6.6	8.6	9.9
MED	5.3	6.5	3.5	6.9	6.5	9	10.6
MIS	6.3	5	4	9.4	7.1	8.7	10.2
MIT	6.7	3	3	9.9	6.9	8.8	10.2
OAK	5.5	6	6	7.3	6.8	8.7	9.5
SAN	7.1	6	8	9	7	8.8	10.7
TOB	7.3	6	6	8.9	6.9	8.2	11.1
TOM	6.3	7	6	8.5	7	8.5	10.7
TOR	7.2	6	8	8	6.9	8	11.5

Note: Product codes and full description of the attributes can be found in chapter 1.

data are decomposed into principal components or factors. These factors are linear combinations of the original predictive variables and are calculated so that a maximum amount of the variation in the x variables (i.e., sensory attributes) is explained. The factors are also orthogonal, meaning that the factor scores for the products have a correlation equal to zero (i.e., the factors are independent from one another). In this particular technique, one usually starts by normalizing the data (Meullenet et al., 2002). Normalizing means that the data for each x variable are centered and the individual observations weighted by the standard deviation across products. The normalized data corresponding to Table 4.1 are given in Table 4.2. For each attribute, the mean across products is zero and the standard deviation is one. This is done so that each attribute is given the same chance to influence the predictive model. This means that x variables of small intensities or small variance across products are given the chance to contribute to the prediction of overall liking. It seems appropriate to normalize the data because there is no *a priori* reason why an attribute with a small intensity or small variance would contribute less to liking for a product than an attribute that has a larger intensity or variance across products.

Most statistical software implementing PLSR will allow data normalization, so this does not have to be done before hand. Features implemented in PLSR usually include the ability to validate the model (e.g., cross validation) and to select x variables that are significantly contributing to the model. This is usually done using a resampling method such as jackknifing. With full cross-validation, each observation (product) is removed one at a time from the sample set, a new model calculation performed, and a predicted score calculated for the sample removed. This procedure is repeated until all samples have been removed from the sample set once. The predictive models can also be optimized using the jackknifing method, which is available as an option with Unscrambler, a multivariate statistical software. Jackknifing is a procedure that was designed to test the significance or lack of significance of the model parameters and that is performed during cross-validation. The root mean square error of prediction (RMSEP), the root mean square error of calibration (RMSEC), and the calibration (R^2) and validation (R^2_{val}) coefficients of determination are usually computed to assess model quality.

PLSR was applied to the tortilla chip data set (see section 1.1), and results of this analysis are presented in Figures 4.1 to 4.5. Figure 4.1 presents the weighted regression coefficients (WRC) for the x variables selected as being significant in the PLSR model

Table 4.2. Mean consumer acceptance and mean normalized descriptive profiles.

Product Codes	Overall Liking	Visual Attributes		Flavor Attributes		Texture Attributes	
		Degree of Whiteness	Grain Flecks	Salt	Grain Complex	Hardness	Crispness
BYW	6.4	0.3	1.2	0.5	0.0	0.2	0.0
GMG	5.5	−2.2	−1.3	−1.2	1.5	2.2	−1.3
GUY	5.9	0.9	0.2	−0.6	−1.4	−0.2	−0.8
MED	5.3	0.6	−1.1	−1.5	−1.9	0.7	0.3
MIS	6.3	−0.3	−0.8	1.1	1.0	0.0	−0.3
MIT	6.7	−1.6	−1.3	1.6	0.0	0.2	−0.3
OAK	5.5	0.3	0.2	−1.1	−0.4	0.0	−1.5
SAN	7.1	0.3	1.2	0.7	0.5	0.2	0.5
TOB	7.3	0.3	0.2	0.6	0.0	−1.2	1.2
TOM	6.3	0.9	0.2	0.2	0.5	−0.5	0.5
TOR	7.2	0.3	1.2	−0.3	0.0	−1.7	1.8

Note: Product codes and full description of the attributes can be found in chapter 1.

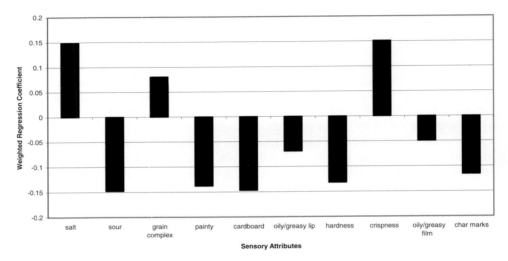

Figure 4.1. Weighted regression coefficients for attributes predicting overall liking of tortilla chips.

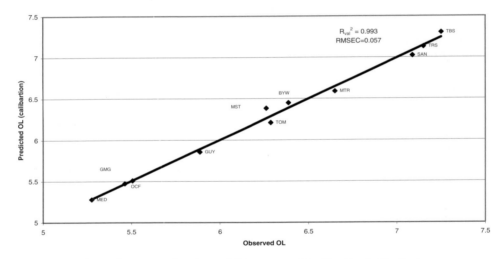

Figure 4.2. Observed versus predicted overall liking scores of tortilla chips (calibration).

(i.e., using jackknifing). The WRCs are not used to actually perform predictions but are used instead to establish the role the x variables play in the prediction of overall liking (OL). A positive WRC signifies that as the corresponding x variable intensity increases, so does OL. This of course assumes that all other x variables in the model are constant. The x variables associated with positive WRCs are usually called positive drivers of acceptance. Correspondingly, x variables with negative WRCs are usually referred to as negative drivers of OL.

In the case of the tortilla chip data, the two main positive drivers of OL were salt and crispness. Negative drivers included sour, painty, cardboard, hardness, and char marks. We can conclude from this that consumers liked those salty and crispy chips devoid of oxidation notes that were not hard and that showed no char marks. Figure 4.2 gives the observed versus predicted liking scores using the PLSR model.

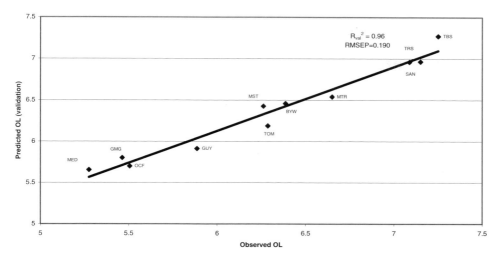

Figure 4.3. Observed versus predicted overall liking scores of tortilla chips (full cross validation).

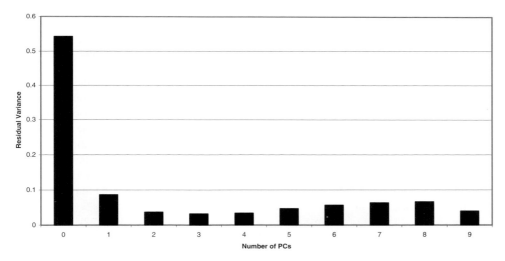

Figure 4.4. Residual validation variance as a function of the number of factors used in the model.

It is evident that the sensory attributes selected by jackknifing for predicting OL did an acceptable job, as the coefficient of determination between observed and predicted values was greater than 0.99. This is, in fact, a fairly unusual result, as R^2 are often quite a bit lower than that reported here. The RMSEC is a measure of the average prediction error expressed in the original OL units. The RMSEC reported is also very low. Figure 4.3 is similar to Figure 4.2 except that predicted OL scores come from the validation step of the analysis. For Figure 4.3, the predicted values were obtained by removing one of the products from the sample set (which is also known as leave-one-out validation), building a PLSR model, and predicting the OL of the sample left out. This is repeated until all observations (i.e., products) have been left out once. This is a useful way to determine whether the model is at all capable of predicting the OL of products not included in the

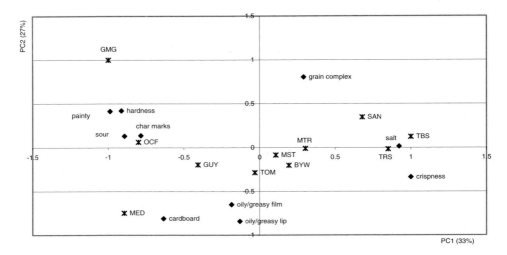

Figure 4.5. Biplot of product scores and variable loading for the model predicting overall liking of tortilla chips.

model. This would become important if the end goal of the analysis is to design new prototypes, profile them sensorially, and attempt the prediction of OL. If the model developed is of good enough quality, this could assist the product developer in refining the sensory properties of a product, with the goal of optimizing OL.

Figure 4.3 shows that the model indeed validated quite well. Although the coefficient of determination was lower and the RMSEP larger than that for the calibration model, the RMSEP is still small enough to endeavor predicting unknown OL values of prototype products not tested by consumers. A useful index to determine whether the RMSEP is small enough is to calculate the discrimination index, referred to as RPD (ratio of prediction to deviation). RPD is the ratio of OL standard deviation across products to RMSEP. A ratio greater than 2 usually indicates that the model can be used for predictions. In our case, the standard deviation for OL was calculated to be 0.70 and the RPD to be 3.68. We therefore conclude that the model can be used for optimization.

Other results that are of interest in this type of analysis include the number of factors best suited for the model. In any regression analysis, there is a balance between the number of predictors necessary to obtain adequate prediction and the risk of overfitting if too many predictors are used. In PLSR, the optimal number of factors used can be estimated by examining a plot of residual validation variance versus the number of factors used. This is illustrated in Figure 4.4. The figure shows that the residual variance decreases up to three factors and then starts to increase with the addition of more factors. In this case, the software used to perform the analysis (Unscrambler, version 9.0) recommended the use of two principal components because the third factor explained very little of the variance.

In addition, biplots of product scores and variable loadings can be useful in examining the relationship between products and variables. This is similar in concept to interpreting biplots from a principal component analysis. In Figure 4.5, the relationship between product and the x variables can be examined. Attributes that are close together on this plot, such as oily/greasy film and oily/greasy lips, were most highly correlated to each other. Products that are closest to specific attributes had higher intensities for this attribute. For example, SAN (Santita's) loaded close to salt and was one of the samples with the highest salt intensity. In summary, PLSR can be a useful method to better understand the drivers of acceptance. However, this technique does not directly provide the sensory profile of the product that will maximize liking by consumers. An alternative to PLSR is nonlinear PLSR. The main disadvantage of linear PLSR is that it is assumed that the relationship between liking and sensory intensity for a given attribute is linear. It is well known that this is not the case for most attributes and that the relationship is usually an inverted U shape. To model the inverted U shape, a quadratic model can be used. This is the general idea behind the nonlinear PLSR we are proposing here. The only difference is that a quadratic term is added to the model for each of the predictors (i.e., sensory variables). From this analysis, the optimal sensory intensity for each of the attributes selected as drivers of OL can be determined. However, this is probably not an optimal modeling procedure, as the correlation among sensory attributes is ignored.

4.2.2. Response Surface Model for External Mapping

External preference mapping refers to a method that uses both consumer and descriptive (i.e., external) data to derive an understanding of sensory dimensions driving liking for a product. The method named PREFMAP, originally proposed by Carroll (1972) and described in Chapter 3, has been modified over the years so that an optimal sensory profile can be derived. We describe here several different approaches and compare the end results. Unfortunately, it is not yet possible to let the reader know what method is best as a comparative study, and validation of the methods is not available in the literature. These methods may yield different optimal product solutions, but unless the products were to be formulated and tested for hedonic level with consumers, there is not a good way to assess superiority of a method over another.

Let us first start with the basic principle of the analysis. The sensory or external data are used to create a multidimensional representation of the sensory stimuli. The analysis can be performed using principal component analysis (PCA). In the case of PCA, singular value decomposition is performed on the mean centered or normalized sensory data. The resulting PCA map is referred to as the sensory space. According to the method described by Danzart (1998), the consumer data for individual consumers are regressed onto the product scores on the first two principal components of the sensory space. The regression model can be expressed as follows:

$$OL_n = \alpha + \sum_{i=1}^{n} \beta_i X_i + \sum_{i=1}^{n} \delta_i X_i^2 + \sum_{i=1}^{n}\sum_{j=1}^{n} \gamma_{ij} X_i X_j, \qquad \text{(Eq. 4.1)}$$

where OL_k is the overall liking score for consumer k, X_i is the product score on PC_i, and α, β, and γ are regression coefficients for the model. The number of principal components used is usually limited to two, as each parameter in the model uses one degree of freedom, and the number of degrees of freedom available for regression is equal to the number of observations minus one (i.e., the number of products in the study).

Let us now examine a specific example. The sensory space was first defined using the average descriptive sensory profiles for the 11 tortilla chip products studied, using a PCA on standardized data. The resulting product scores are given in Figure 4.6, whereas the attribute loadings are given in Figure 4.7. The product scores presented in Figure 4.6 are those that are used as explanatory variables (i.e., x variables) in Equation 4.1. The product scores represent the coordinates of the products in the sensory space. Products that are close on the map are similar from a sensory standpoint, whereas the products that

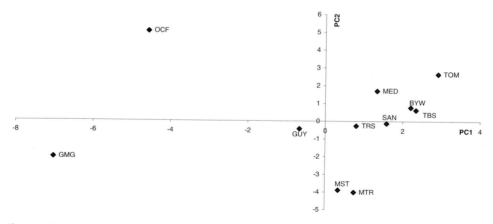

Figure 4.6. Product scores on first two principal components.

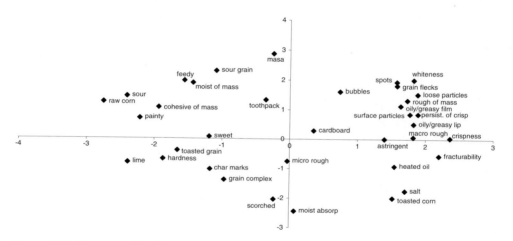

Figure 4.7. Sensory attribute loadings on the first two axes.

are most distant on the map are most dissimilar overall. For example, products MST and MTR had very similar coordinates in the sensory space. This is not surprising because these products are made by the same company, and the only difference between the two products is their shape, one being a strip and the other a triangle. In addition, Figure 4.6 indicates that OCF and GMG were most dissimilar to the other products.

Figure 4.7 presents the attribute loadings on the first two axes of the sensory space. This can be useful for examining the relationship among attributes and determining the reasons for product loadings in the sensory space.

For each consumer, the individual's liking scores for the products are used as the y variable, and the model in Equation 4.1 is fitted using as x variables the product scores on the first two axes of the sensory space (Table 4.3). The resulting surface responses for three consumers who participated in the tortilla chip consumer test are presented in Figure 4.8. The x and y axes represent the sensory space, whereas the z axis represents OL. These three individuals were selected because they each represent a different type of consumer.

Table 4.3. Estimates of response surface models (OL = $b_0 + b_1 \times PC_1 + b_2 \times PC_2 + b_3 \times PC_1 \times PC_1 + b_4 \times PC_1 \times PC_2 + b_5 \times PC_2 \times PC_2$).

Consumer ID	R^2	b_0	b_1	b_2	b_3	b_4	b_5
1	0.33	4.29	1.03	−0.34	0.15	−0.13	0.00
2	0.35	5.36	0.07	0.33	−0.04	0.03	0.11
3	0.49	6.65	0.72	−0.64	0.19	−0.35	−0.21
4	0.54	5.20	0.40	−0.22	0.08	−0.02	0.14
5	0.32	7.01	0.02	0.12	0.02	−0.07	−0.14
6	0.61	4.53	0.95	−0.56	0.08	−0.15	0.03
7	0.47	5.83	0.45	−0.61	0.12	−0.22	−0.13
8	0.75	4.69	0.90	−0.43	0.20	−0.26	−0.06
9	0.71	7.42	0.00	−0.24	0.05	−0.08	0.00
10	0.55	6.73	0.25	−0.22	0.05	−0.04	−0.02
11	0.53	7.15	−0.25	−0.05	−0.11	0.14	0.14
12	0.62	6.45	0.32	−0.08	0.02	0.07	−0.09
13	0.80	4.93	0.83	−0.66	0.14	−0.21	−0.06
14	0.62	4.32	0.18	−0.20	0.07	0.05	0.13
15	0.54	3.06	1.39	−0.84	0.30	−0.20	0.03
16	0.62	7.95	−0.23	0.21	−0.06	0.02	−0.09
17	0.45	4.96	0.80	−0.58	0.21	−0.19	0.00
18	0.67	7.55	0.03	0.18	−0.08	0.02	−0.04
19	0.76	4.78	0.81	−0.21	0.16	−0.01	0.09
20	0.66	6.17	0.24	−0.71	0.12	−0.23	−0.13
21	0.54	5.05	0.92	−0.35	0.16	−0.18	0.01
22	0.25	6.34	0.20	−0.07	0.02	−0.07	−0.08
23	0.24	6.44	0.22	−0.36	0.08	−0.07	−0.12
24	0.64	7.40	0.13	−0.13	−0.01	−0.08	−0.04
25	0.80	6.74	0.25	−0.24	0.00	0.07	−0.07
26	0.41	6.72	0.52	−0.09	0.14	−0.16	−0.13
27	0.30	8.70	−0.26	0.02	−0.02	−0.02	−0.04
28	0.85	8.80	−0.87	0.33	−0.10	−0.24	−0.39
29	0.76	7.80	0.24	0.05	0.04	−0.11	−0.09
30	0.66	5.79	0.83	−0.34	0.18	−0.20	−0.22
31	0.46	4.33	0.64	−0.47	0.07	−0.03	0.06
32	0.87	5.98	1.08	−0.24	0.11	−0.06	−0.10
33	0.33	7.25	−0.02	−0.19	−0.01	−0.18	−0.21
34	0.73	5.66	0.16	0.32	0.02	−0.21	−0.15
35	0.87	7.15	0.56	0.24	0.04	−0.15	−0.11
36	0.32	7.00	0.34	−0.39	0.09	−0.17	−0.10
37	0.68	5.45	−0.40	−0.22	−0.15	0.08	0.07
38	0.70	4.43	0.62	−0.73	0.16	−0.14	−0.10
39	0.68	6.90	0.17	0.01	0.05	−0.08	−0.04
40	0.59	6.21	0.56	−0.43	0.13	−0.27	−0.18
41	0.93	6.24	0.85	−0.44	0.16	−0.07	−0.04
42	0.89	7.30	0.59	−0.88	0.10	−0.38	−0.39
43	0.65	6.98	0.25	0.06	−0.10	0.09	−0.04
44	0.93	7.27	0.13	−0.22	−0.01	−0.14	−0.10
45	0.84	5.46	0.92	−0.06	0.06	−0.05	0.14
46	0.95	4.19	0.33	−0.32	0.08	−0.03	0.02
47	0.36	6.11	0.97	−0.65	0.24	−0.28	−0.23
48	0.72	6.53	0.63	0.15	0.08	−0.16	−0.07
49	0.19	5.38	0.27	−0.11	0.06	−0.10	−0.01
50	0.34	5.35	1.07	−0.52	0.28	−0.37	−0.21
51	0.66	5.89	0.72	−0.28	0.20	−0.40	−0.30
52	0.56	3.23	1.30	−0.61	0.30	−0.26	−0.02
53	0.67	4.80	0.95	−0.27	0.25	−0.17	−0.09

(*Continued*)

77

Table 4.3. Continued.

Consumer ID	R^2	b_0	b_1	b_2	b_3	b_4	b_5
54	0.29	5.03	0.40	−0.35	0.14	−0.23	−0.10
55	0.51	7.10	0.55	−0.20	0.13	−0.22	−0.17
56	0.37	6.50	−0.25	−0.02	−0.02	0.06	0.01
57	0.29	6.86	−0.35	0.00	−0.09	0.06	−0.01
58	0.33	8.16	−0.52	0.30	−0.12	0.12	0.01
59	0.23	4.76	0.55	−0.08	0.15	−0.06	0.03
60	0.82	7.87	−0.21	−0.11	0.02	−0.08	−0.06
61	0.53	7.35	0.27	−0.01	0.03	−0.06	−0.04
62	0.41	7.34	−0.05	−0.08	0.00	−0.09	−0.09
63	0.94	4.71	1.18	−0.61	0.23	−0.14	−0.01
64	0.39	5.85	0.73	−0.32	0.08	−0.18	−0.11
65	0.88	7.59	0.29	0.02	−0.02	0.00	−0.07
66	0.42	5.93	0.22	−0.50	0.04	−0.20	−0.23
67	0.40	6.20	0.56	−0.18	0.10	−0.16	−0.07
68	0.67	3.67	1.19	−1.10	0.34	−0.39	−0.15
69	0.51	4.70	0.98	−0.58	0.20	−0.39	−0.25
70	0.88	6.97	0.57	−0.33	0.09	−0.12	−0.06
71	0.51	7.54	−0.23	0.34	−0.04	0.11	0.04
72	0.57	6.59	−0.01	−0.18	0.01	−0.04	0.01
73	0.54	8.01	0.11	−0.11	0.00	−0.03	−0.03
74	0.68	5.63	0.93	−0.61	0.09	−0.22	−0.06
75	0.79	4.77	1.34	−1.20	0.35	−0.53	−0.35
76	0.49	2.53	1.14	−0.70	0.25	−0.12	0.09
77	0.65	5.92	0.35	−0.78	0.07	−0.21	−0.11
78	0.33	5.67	0.58	−0.05	0.06	−0.06	0.05
79	0.53	3.53	0.31	0.45	0.06	0.01	0.10
80	0.65	4.96	0.21	0.11	−0.03	0.13	0.04

Consumer 16 represents a consumer who clearly expressed a preference for one specific region of the sensory space—the optimal OL response is predicted to be 8.24 and has coordinates of −1.67 and 0.89 on the first and second sensory space axes, respectively.

Consumer 14 represents a consumer who expressed a clear dislike for one region of the sensory space. In this case, the minimum predicted OL score of 4.04 was found to have coordinates of −1.66 and 1.11 on the first and second sensory space axes, respectively.

Finally, consumer 12 represents an eclectic consumer who did not exhibit a clear preference for one specific small area of the sensory space. The examination of the response surface indicates that this consumer's liking of the products was not greatly affected by the products' coordinates on the first sensory space axis. However, coordinates on axis two seemed to have a more profound effect on this subject's liking for the products. Danzart (2004) found this to be a very common occurrence, with as many as 75% of consumers being classified as eclectics in various studies.

The next step in this analysis is to define for each individual what the acceptable portion of the sensory space is for each consumer. This is easily performed by estimating the areas of the sensory space that are acceptable to this consumer. First, one needs to identify what minimum OL score is acceptable for an individual consumer. We believe it is necessary to establish an acceptable score for individual respondents because, as with many other scales used in sensory and consumer sciences, not every individual will use the scale in a similar manner. Danzart et al. (2004) propose to use the mean consumer OL score across products for the individual as the acceptable OL level for the individual. Although we

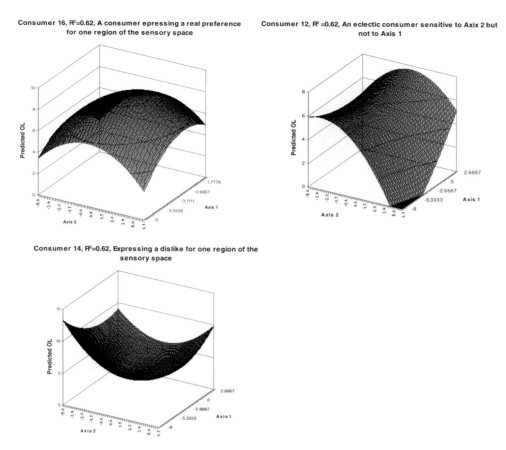

Figure 4.8. Types of consumer responses to the sensory space.

feel this is too lenient of a rule, we apply it here because our description of this method is almost entirely inspired by the work published by Danzart.

Figure 4.9 shows the acceptable regions of the sensory space for consumers 12, 14, and 16. For consumer 12, the eclectic, the acceptable region (OL > 6.0) was found for coordinates greater than −2 on PC1 and between −3 and 4 for PC2. For consumer 16, an individual expressing a clear preference, the acceptable area of the sensory space was elliptical, as seen in Figure 4.9. Finally, for consumer 14, who expressed a clear dislike for the center of the sensory space, the acceptable region is at the periphery of the space. These maps clearly illustrate the lack of homogeneity of the consumer population. However, it is the ability to map out differences among consumers that is most appealing about this method.

The next step in the analysis is to compile the information obtained for individuals into one final map. This is done by superimposing all the individual contour plots showing the acceptable regions of the sensory space. This method is, overall, a very similar approach to the surface response methodology when multiple responses are used to determine the optimal level of two or more factors. In practice, the acceptable region for an individual consumer is assigned a value of one, and the unacceptable region is given a value of zero. The result of the superimposition is given in Figure 4.10. The various contours in the map represent the percentage of consumers liking a virtual product located in a particular

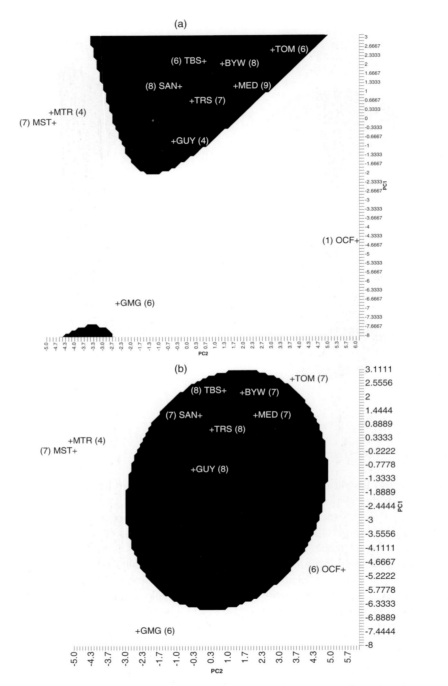

Figure 4.9. Contour plots showing the acceptable region of the sensory space for consumers 12, 14, and 16. Acceptable region in black. Acceptable region determined for predicted scores above the mean (X = 6.0 for consumer 12; X = 5.7 for consumer 14; X = 6.8 for consumer 16).

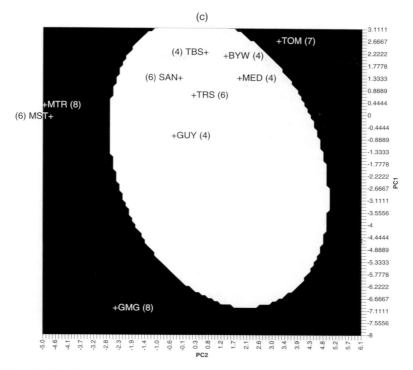

Figure 4.9. *Continued.*

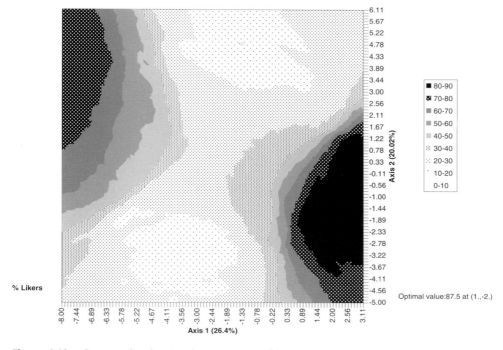

Figure 4.10. Contour plot showing the percentage of likers in the regions of the sensory space. This map is obtained by superimposing acceptance areas for individual consumers.

position on the sensory map. The map also reports the coordinates of the point for which the percentage of likers is maximum. This value was found to 87.5% of likers at coordinates of 1 and −2 on axes 1 and 2, respectively. The map also shows that products scoring above 1 on axis 1 and between −3.5 and 1.2 on axis 2 would satisfy a high proportion of consumers (80%–90%). Alternatively, products scoring between −8.0 and −6.0 on axis 1 and between 0.8 and 6.0 on axis 2 would also satisfy a high percentage of consumers (70%–80%). This signifies that this type of analysis may yield multiple optimum points in the sensory space and that other considerations such as avoiding "me too" products could be taken into account to determine an optimal product.

Unfortunately, the location in the sensory space does not tell the experimenter what the sensory profile of the optimal product is. This could be done by regression, but Danzart has proposed a barycenter method that is capable of determining the profile of the optimal product on the basis of three, carefully chosen products included in the original consumer test (Danzart et al., 2004). The coordinates of the optimal product can be estimated as the linear combination of the coordinates of three products close to the optimal product. The barycenter method finds the three nearest products (A, B, and C) to the optimal product (P) on the sensory space and then calculates the weights (a, b, c) that define P as the barycenter of $\{(A,a), (B,b), (C,c)\}$ with the constraint $a + b + c = 1$. On the basis of the calculated values of a, b, and c, the sensory profile of the optimal (P) can be determined as a weighted average of products A, B, and C's sensory profiles (Blumenthal, 2004). The choice of the products for our tortilla chip example is given in Figure 4.11.

In our example, TRS, SAN, and MTR were selected as the three products to assess the sensory profile of the optimum product. The coordinates of these products in the sensory space were (0.80, −0.26), (1.58, −0.12), and (0.75, −4.02) for TRS, SAN, and MTR, respectively, whereas (1.0, −2.0) were the coordinates of the optimal product. We now have the following set of equations that need to be satisfied:

$$aX_{1TRS} + bX_{1SAN} + cX_{1MTR} = X_{1OPT};$$

$$aX_{2TRS} + bX_{2SAN} + cX_{2MTR} = X_{2OPT};$$

$$a + b + c = 1, \qquad\qquad \text{(Eq. 4.2)}$$

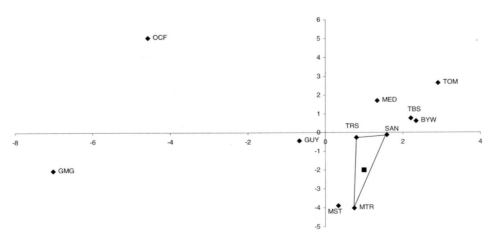

Figure 4.11. Coordinates in the sensory space of the optimal product in comparison to the products tested.

where X_1 is the coordinates on the first axis of the sensory space, X_2 is the coordinates on the second axis, and a, b, and c are the barycenter weights. Solving for a, b, and c gives values of 0.2408, 0.286, and 0.4732, respectively. The following equation is then used to obtain the sensory profile of the optimum product:

$$S_{kOPT} = 0.2408 S_{kTRS} + 0.286 S_{kSAN} + 0.4732 S_{kMTR}, \quad \text{(Eq. 4.3)}$$

where S_k is the intensity for attribute k. In our example, the barycenter method is appropriate because we could find three products that formed a triangle, and the optimum product location was inside that triangle. However, this is a special case, and there could be situations in which the optimal product is outside of any triangle that can be formed. In this situation, one of the weights will be negative, and the reliability of the optimal product is—in our estimation—greatly diminished. In addition, the choice of the three products will affect the results of the calculation of the optimal product. Furthermore, this method is appropriate for only two axes because the optimal product on the PC1–PC2 sensory space may not be in the maximal preference areas on the other dimensions of the sensory space (PC1–PC3, PC1–PC4, PC2–PC3, etc.). We therefore conclude that the barycenter method to determine the sensory profile of the optimum product has limitations, and we propose here an alternative method.

We propose using the generalized inverse matrix to obtain the optimal sensory profile. Because this method has not been published, we give here the theoretical thought involved in the development of this method.

Response surface model (RSM) is widely used for optimization purposes and we have relied so far on the use of this methodology. In the proposed approach, it is used to model the relationship between principal components (PCs) and the response (OL). The expression is given here,

$$OL = RSM(PC_1, PC_2, \ldots, PC_m), \quad \text{(Eq. 4.4)}$$

where m is the number of PCs in the RSM, which can be determined by the analyst. Note that we are not limited here to two axes to define the sensory space.

Through an optimization process, we can have the maximum response OL_{max} at the optimal point,

$$(PC_1^*, PC_2^*, \ldots, PC_m^*); \quad \text{(Eq. 4.5)}$$

namely,

$$OL_{max} = RSM(PC_1^*, PC_2^*, \ldots, PC_m^*). \quad \text{(Eq. 4.6)}$$

According the definition of principal components, the optimal point,

$$(PC_1^*, PC_2^*, \ldots, PC_m^*), \quad \text{(Eq. 4.7)}$$

has the following equations with the transformed sensory variables $(Z_1, Z_2, \ldots, Z_n)$:

$$PC_1^* = a_{11}Z_1 + a_{12}Z_2 + \ldots + a_{1n}Z_n;$$
$$PC_2^* = a_{21}Z_1 + a_{22}Z_2 + \ldots + a_{2n}Z_n;$$
$$\ldots$$
$$PC_m^* = a_{m1}Z_1 + a_{m2}Z_2 + \ldots + a_{mn}Z_n. \quad \text{(Eq. 4.8)}$$

In the generation of the sensory space by principal component analysis, there are three options for the transformed variables Z_i ($i = 1, 2, \ldots, n$): first, using the original variables $Z_i = X_i$; second, using centered variables $Z_i = X_i - \overline{X}_i$; and third, using standardized variables $Z_i = \dfrac{X_i - \overline{X}_i}{\sigma_i}$, where $\overline{X}_i$ and σ_i are the mean and standard deviation of X_i (with X_i representing the sensory attributes used in the analysis). In terms of matrix notation, it can be reexpressed as

$$AZ = B, \qquad \text{(Eq. 4.9)}$$

where

$$A = \begin{pmatrix} a_{11} & a_{12} & \cdots & a_{1n} \\ a_{21} & a_{22} & \cdots & a_{2n} \\ \cdots & \cdots & \cdots & \cdots \\ a_{m1} & a_{m2} & \cdots & a_{mn} \end{pmatrix};$$

$$Z = \begin{pmatrix} Z_1 \\ Z_2 \\ \cdots \\ Z_n \end{pmatrix};$$

and

$$B = \begin{pmatrix} PC_1^* \\ PC_2^* \\ \cdots \\ PC_m^* \end{pmatrix}. \qquad \text{(Eq. 4.10)}$$

There are three possible scenarios for the dimensionality of these matrices: first, $m = n$; second, $m < n$; and third, $m > n$. If $m = n$, A is a square matrix. The system $AZ = B$ has a unique solution if and only if A is nonsingular or if the inverse of A (denoted by A^{-1}) exists. Numerous algorithms can be used to solve the system; namely, $Z = A^{-1}B$. However, $m = n$ is usually not true for most cases of preference maps. When $m \neq n$, A is no longer a square matrix and the inverse of A cannot be calculated. It also means that there are no unique solutions to the system $AZ = B$ when $m \neq n$. However, a generalized inverse of A can be used to solve the system $AZ = B$, and the solution is $Z = A^- B$ (A^- is a generalized inverse). If the system $AZ = B$ is consistent, then the solution $Z = A^- B$ is unique if and only if $A^- A = I$. Let A be an $M \times N$ matrix. If a matrix $A^\dagger$ exists that satisfies the four conditions below, we shall call $A^\dagger$ a Moore-Penrose or Penrose inverse of A:

$$AA^\dagger = (AA^\dagger)^T;$$
$$A^\dagger A = (A^\dagger A)^T;$$
$$AA^\dagger A = A;$$
$$A^\dagger AA^\dagger = A^\dagger. \qquad \text{(Eq. 4.11)}$$

For each matrix A, there exists a unique matrix $A^\dagger$ that satisfies the Moore-Penrose conditions. That is, each matrix A has a unique Penrose inverse $A^\dagger$. This indicates that the solution to $AZ = B$ is unique, too.

The Singular Value Decomposition method is often used to calculate a generalized inverse matrix $A^\dagger$, and the solution is then obtained; namely,

$$Z^* = A^\dagger B = \begin{pmatrix} Z_1^* \\ Z_2^* \\ \ldots \\ Z_n^* \end{pmatrix}. \quad \text{(Eq. 4.12)}$$

Having found the optimal transformed sensory variable matrix Z^*, the optimal sensory profile (X^*) of the product can be calculated as follows:

$$X^* = \begin{pmatrix} X_1^* \\ X_2^* \\ \ldots \\ X_n^* \end{pmatrix} = \begin{pmatrix} Z_1^* \\ Z_2^* \\ \ldots \\ Z_n^* \end{pmatrix} \quad \text{if using the original variables;}$$

$$X^* = \begin{pmatrix} X_1^* \\ X_2^* \\ \ldots \\ X_n^* \end{pmatrix} = \begin{pmatrix} \bar{X}_1 + Z_1^* \\ \bar{X}_2 + Z_2^* \\ \ldots \\ \bar{X}_n + Z_n^* \end{pmatrix} \quad \text{if using the centered variables; and}$$

$$X^* = \begin{pmatrix} X_1^* \\ X_2^* \\ \ldots \\ X_n^* \end{pmatrix} = \begin{pmatrix} \bar{X}_1 + Z_1^* \sigma_1 \\ \bar{X}_2 + Z_2^* \sigma_2 \\ \ldots \\ \bar{X}_n + Z_n^* \sigma_n \end{pmatrix} \quad \text{if using the standardized variables.} \quad \text{(Eq. 4.13)}$$

Let us now take an example and compare the results given by the barycenter and g-inverse methods. The profile of the optimum product determined for the tortilla chip data set using either the barycenter or g-inverse method is given in Table 4.4. Overall, the two methods are in close agreement with each other, with only a few attributes differing from a practical sense of view. For example, the degree of whiteness was predicted to be 5.44 for the g-inverse method and only 4.58 for the barycenter method. Similarly, char marks was predicted to be 3.5 by the g-inverse and only 2.56 by the barycenter. Finally, the optimal salt level was predicted to be 8.41 by the g-inverse and 9.18 by the barycenter. Overall, these discrepancies are rather small. One advantage of the g-inverse method is that the method does not require the selection of specific products to perform the analysis. In fact, the reliance of the barycenter method on an adequate product choice causes some significant variations in the prediction of the optimal product, as illustrated in Table 4.4. To illustrate the variability of the type of answers provided by the barycenter method, calculations were performed using TRS, SAN, and MTR; GUY, BYW, and MST; or MED,

Table 4.4. Optimal sensory profile determined by g-inverse and barycenter methods.

		Barycenter Method		
	g-Inverse	MTR/SAN/TRS	GUY/BYW/MST	MED/TBS/MST
Visual				
Degree of whiteness	5.66	4.58	5.46	5.45*
Grain flecks	6.10	5.63	5.58	4.96*
Char marks	3.17	2.56	3.96	3.41*
Surface particles	2.16	1.71	2.25	2.44
Amount of bubbles	6.00	5.57	5.42	5.79
Spots	9.92	9.20	9.14	8.70*
Flavor				
Sweet	0.49	0.53	0.56	0.54
Salt	8.98	9.18	9.12	8.99
Sour	0.01	0.02	0.07	0.07
Lime	−0.01	0.00	0.00	0.00
Astringent	2.58	2.55	2.61	2.62
Grain complex	6.89	6.94	6.99	6.94
Toasted corn	3.01	2.95	3.31	3.31*
Raw corn	−0.03	0.00	0.00	0.00
Masa	3.53	3.37	3.37	3.44
Toasted grain	1.44	1.71	1.35	1.20*
Painty	0.01	0.00	0.00	0.02
Feedy	−0.03	0.00	0.00	0.00
Heated oil	4.45	4.34	4.46	4.48
Scorched	0.63	0.26	1.62	1.73*
Cardboard	2.47	2.38	2.59	2.55
Sour grain	−0.03	0.00	0.00	0.00
Texture				
Micro roughness	8.06	8.15	8.37	8.05
Macro roughness	4.22	4.26	3.73	4.08
Oily/greasy lip	6.10	5.60	6.46	6.19*
Loose particles	5.76	5.68	5.79	5.50
Hardness	8.56	8.58	8.74	8.60
Crispness	10.72	10.64	10.27	10.52
Fracturability	7.78	7.77	7.48	7.67
Cohesive of mass	2.98	3.09	2.82	2.98
Roughness of mass	7.49	7.53	7.41	7.29
Moistness of mass	7.19	7.10	7.30	7.28
Moisture absorption	9.40	9.56	9.42	9.43
Persistence of crisp	5.47	5.41	5.34	5.22
Toothpack	5.16	5.03	5.30	5.34
Oily/greasy film	4.01	3.88	3.90	3.92

*Indicates attributes for which differences in optimal solutions were greater than the panel Least Square Difference ($\alpha = 0.05$).

TBS, and MST. It is clear that for the three examples given, the method does a decent job at giving agreeing answers. However, there were a few discrepancies. For example, the optimal level for degree of whiteness was 4.56 when TRS/SAN/MTR were used and 5.46 and 5.45 when GUY/BYW/MST and MED/TBS/MST were used, respectively.

On the basis of the description of the method proposed by Danzart, this seems to be a very robust method and is probably sufficient to answer most product optimization problems in the food industry. However, the method is not perfect, and we here point out some of its weaknesses. One of the issues with the method concerns the fit (or lack

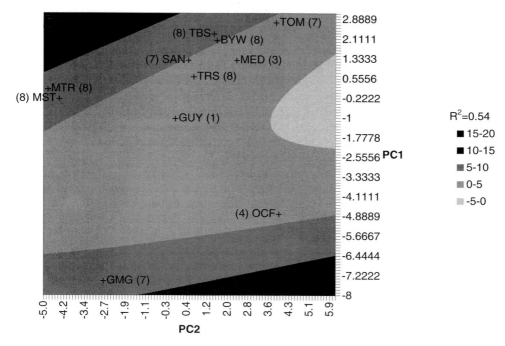

Figure 4.12. Example of a response surface model resulting in poor prediction of overall liking scores (consumer 15).

of) of response surfaces to data for individual consumers. In many cases, the R^2 values for some of these responses are rather low and often translate into poor predictions by the model of the actual hedonic data. This can in turn lead to an inaccurate assessment of the acceptable region of the sensory space for a particular consumer. This is illustrated in Figure 4.12 for consumer 15. The model fit was acceptable, with $R^2 = 0.54$, and for the most part the products evaluated fell in the correct contour band. However, some of the predicted values in the map were above 15, which is nonsensical because the hedonic scale varies from 1 to 9. One of the reasons for this error is that in the overpredicted regions, we had no product present. In addition, when a RSM is used, 6 degrees of freedom are used, and when the total number of observations is 11, as in our case study, there is potential for data overfitting. Nevertheless, the RSM method is probably the most elegant method available for determining the sensory profile of an optimal product.

4.2.3. Euclidian Distance Approach

We now present an alternative but similar method that does not rely on the use of response surfaces to assess the consumer liking map and to determine the location in the sensory space of the optimal product. Because this method has not been described in the literature, as it is a result of recent research in our laboratories, we first describe its concept before applying it to a sample data set. Let us first consider a sensory space, as depicted in Figure 4.13. This is a representation of a sensory space with product locations (P1 – P10) and with a series of vectors representing individual consumer preferences. This representation is that of a standard external preference map in which consumers are projected in the sensory space. Let us now consider a specific consumer, as depicted in Figure 4.13.

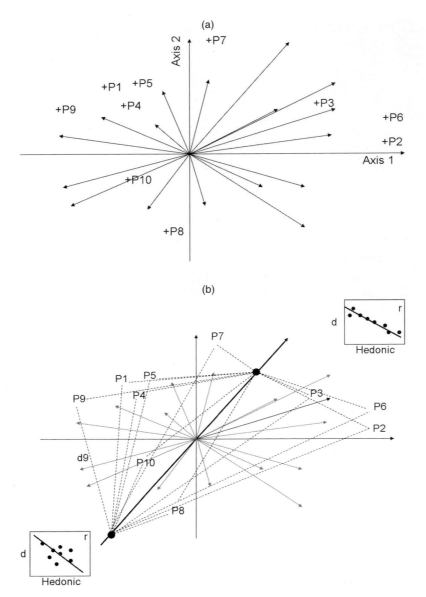

Figure 4.13. Illustration of (a) vectors indicating preference directions for 10 products and (b) concept of Euclidian distances between a point in the space and the products evaluated.

The arrow depicts the direction of preference in the sensory space for that individual. We know that the vector model described in the MDPREF framework is unable to identify the location of an ideal point for an individual. The ideal point in the sensory space should be, in concept, closest to products with a high OL and farthest from products with a low OL. The proximity between two points in the sensory space can be simply assessed using Euclidian distances. Let us now go back to Figure 4.13. We have taken two locations on the consumer vector selected. One is in the upper-right quadrant, which should be closer to the ideal point than the second point, which is in the lower-left quadrant, because of the direction of the preference. The distances between each of the two points and the products can then be calculated and correlated to OL values for our 10 products.

A correlation between the hedonic scores and the distances between the point considered a potential ideal and the products in the sensory space can easily be calculated. One would expect the correlation to be negative, as the ideal should be found close (small distance) to the products with high OL scores. We then expect the correlation to be higher in absolute value for the point in the upper-right quadrant than for the point in the lower-left quadrant of the map. In practice, the sensory space is divided into a fine mesh, and these correlations are assessed at every point in the grid or mesh, as depicted in Figure 4.14. The ideal point for a given consumer is found at the point in the grid where the correlation is minimum (i.e., most highly negative). After the ideal point has been found, correlations for the neighboring points in the grid are statistically compared to the minimum correlation. If the correlation is not statistically significant from the minimum correlation, then this second location is added to the acceptable region of the sensory space for this consumer. This statistical test is repeated for every point in the map, and the acceptable region of the sensory space is determined. As for the Danzart method previously described, the maps are overlaid for all consumers and the final map drawn. The maximum point in the sensory space is then taken as the group ideal or optimal product and its sensory profile assessed, using the previously described g-inverse method or least squares regression.

Let us now apply this method to the tortilla chip data and compare the results to those of the RSM method. Figure 4.15 gives the acceptable region (in black) of the sensory space for consumers 12, 14, and 16. Compared to results from RSM, the acceptable regions of the sensory space are similar for all three examples taken, although their shape is slightly less regular for the Euclidian distance correlation (EDC) method. The superimposition on the maps of the products and their hedonic scores show that for the most part, the products with high hedonic levels were located in the acceptable region of the sensory space.

Figure 4.16 gives the final map for the EDC method. The map was slightly different than that for the RSM method, and accordingly, the optimal point in the sensory space had different coordinates (1.59, −0.06) on axes 1 and 2, respectively.

The optimal sensory profile for the EDC approach was evaluated using the g-inverse method and the profile compared to that given by RSM (Table 4.5). Overall, the results were extremely similar. Some slight differences in intensity were found for degree of whiteness (0.28), grain flecks (0.31), spots (0.41), salt (0.35), and toasted corn (0.23). However, these differences were less than the respective least square differences for the various attributes, and we conclude that the optimal profiles yielded by either the RSM or EDC methods were not statistically different from each other.

The example given here for the EDC method was conducted in the external mapping framework, as the product configuration space was determined using external data (i.e., sensory profile data). However, the same method can be applied to internal maps in which the product configuration is obtained from consumer data. Instead of presenting this analysis here, we present it in Chapter 5 in comparison to a method of unfolding.

4.3. Conclusions

Several papers are quite critical of and underline some limitations of external preference mapping (Blumenthal 2004; Faber et al., 2003; Guinard et al., 2001). In particular, two weaknesses seem to be consistently brought forward. First, many prefer the internal approach, as the product configuration is derived from hedonic data. This results in

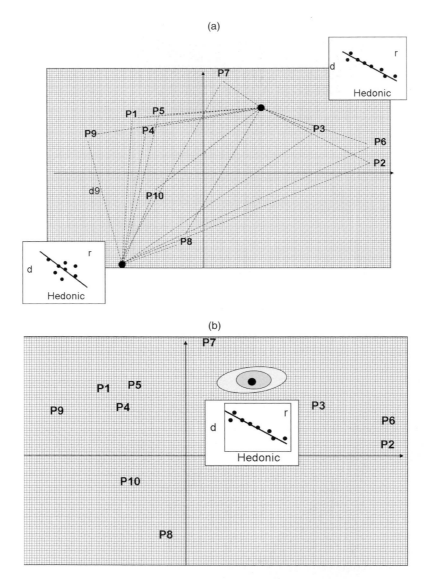

Figure 4.14. Illustration of the Euclidian Distance Correlation method for identifying consumer ideals: (a) distances between two potential ideal points and the 10 products and their correlation to liking scores and (b) acceptance region for a specific consumer defined as the region for which the correlation between distances and hedonic scores are not significantly different from the maximum according to a significance *t*-test described by Snedecor and Cochran (1967) for samples with correlated variances.

products with similar hedonic levels being located close to each other in the space. The critics of external preference mapping argue that arranging products in a sensory space without considering the hedonic data could be misleading, as sensory attributes that drive the location of individual products in the space may not be attributes that have an effect on product liking. This artificial product configuration leads to the second criticism of the method—the poor fit offered by response surface models for a significant proportion of consumers. Faber et al. (2003) discuss this at length and report that between 36% and

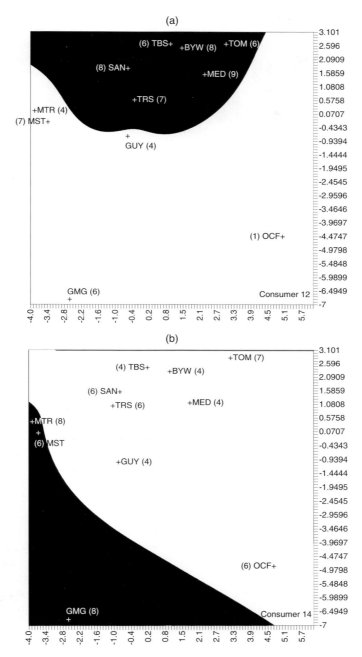

Figure 4.15. Acceptance region using the Euclidian Distance Correlation method with $\alpha = 0.05$ for (a) consumer 12, (b) consumer 14, and (c) consumer 16.

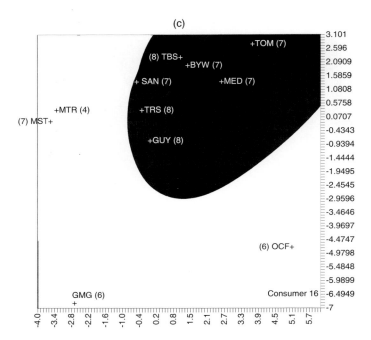

Figure 4.15. *Continued.*

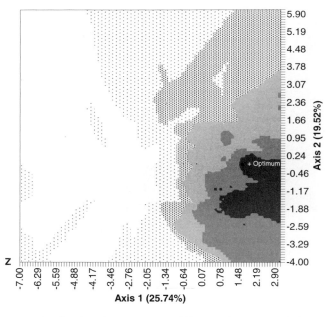

Figure 4.16. Contour plot showing the percentage of likers in the regions of the sensory space using the Euclidian distance correlation approach. This map is obtained by superimposing acceptance areas for individual consumers.

Table 4.5. Optimal sensory profile determined by the g-inverse method for the RSM and Euclidian distance approaches.

	Euclidian Distance	RSM
Visual		
Degree of whiteness	5.94	5.66
Grain flecks	6.41	6.10
Char marks	2.98	3.17
Surface particles	2.16	2.16
Amount of bubbles	6.14	6.00
Spots	10.33	9.92
Flavor		
Sweet	0.50	0.49
Salt	8.63	8.98
Sour	0.05	0.01
Lime	−0.01	−0.01
Astringent	2.57	2.58
Grain complex	6.86	6.89
Toasted corn	2.73	3.01
Raw corn	0.00	−0.03
Masa	3.67	3.53
Toasted grain	1.48	1.44
Painty	0.04	0.01
Feedy	0.00	−0.03
Heated oil	4.42	4.45
Scorched	0.40	0.63
Cardboard	2.48	2.47
Sour grain	0.04	−0.03
Texture		
Micro roughness	8.01	8.06
Macro roughness	4.14	4.22
Oily/greasy lip	6.07	6.10
Loose particles	5.82	5.76
Hardness	8.56	8.56
Crispness	10.64	10.72
Fracturability	7.72	7.78
Cohesive of mass	3.04	2.98
Roughness of mass	7.50	7.49
Moistness of mass	7.28	7.19
Moisture absorption	9.31	9.40
Persistence of crisp	5.47	5.47
Toothpack	5.19	5.16
Oily/greasy film	4.02	4.01

75% of consumers have been reported in the literature to be significantly fitted. Several researchers, including Greenhoff and MacFie (1994), attributed this poor fit to the loss of information in higher-numbered PCs, as external preference mapping commonly employs only the first two PCs (Faber et al., 1993). The other mentioned weakness of extensions of external preference mapping is the fact that with polynomial models such as the quadratic model used by Danzart (1998), 5 degrees of freedom are required to fit the data. When the number of products present in the map is small, overfitting is a relevant issue. We have proposed in this chapter both a new method to determine the location of the ideal point in an external sensory space and an alternative to the barycenter method of determining the sensory profile of the ideal point. EDC mapping could be used as an alternative to the

RSM method proposed by Danzart, as it would solve some of the overfitting problems of quadratic models. The g-inverse solution to determine the optimal sensory profile could be used as an enhancement of RSM, as the barycenter solution is dependent on the arbitrary choice of three products.

References

Blumenthal, D. 2004. How to obtain the sensory scores of the optimal product according to external preference mapping. 6th Sensometric Meeting, Davis, CA USA, August 2004.

Carroll, J.D. 1972. Individual differences and multidimensional scaling. In R.N. Shepard, A.K. Romney, and S.B. Nerlove (Eds.). Multidimensional scaling: theory and applications in the behavioral sciences (pp. 105–155). New York: Seminar Press.

Danzart, M. 1998. Quadratic model in preference mapping. 4th Sensometric meeting, Copenhagen, August 1998.

Danzart, M., Sieffermann, J-M. and Delarue, J. 2004. New developments in preference mapping techniques: Finding out a consumer optimal product, its sensory profile and the key sensory attributes. 6th Sensometric Meeting, Davis, CA USA, August 2004.

Faber, N.M., Mojet, J. and Poelman, A.A.M. 2003. Simple improvement of consumer fit in external preference mapping. Food Quality and Preference 14(2003): 455–461.

Guinard, J.-X., Uotani, B. and Schlich, P. 2001. Internal and external mapping of preferences for commercial lager beers: Comparison of hedonic ratings by consumers blind vs. with knowledge of brand and price. Food Quality and Preference. 12(4):243–255 (2001).

Greenhoff, K. and MacFie, H.J.H. 1994. Preference mapping in practice. In H.J.H. MacFie and D.M.H. Thomson (Eds.) Measurement of food preferences (pp. 137–147). Glasgow: Blackie Academic and Professional.

Meullenet, J.-F., Xiong, R. Monsoor, M. Bellman-Horner, T. Zivanovic, S. Dias, P. Fromm, H. and Liu, Z. 2002. Preference mapping of commercial toasted white corn tortilla chips. J. Food Sci. 67:1950–1957.

Moskowitz, H.R. 1985. New directions in Product testing and sensory analysis of food. Westport: Food and Nutrition Press.

Snedecor, G.W., and W.G. Cochran. 1967. Statistical methods (6th ed.). Ames, IA: Iowa State University Press.

5 Multidimensional Scaling and Unfolding and the Application of Probabilistic Unfolding to Model Preference Data

5.1. Introduction

Multidimensional scaling (MDS) is a technique used to represent similarity or dissimilarity among objects (often products in sensory problems) in a low-dimensional space or map. MDS has its origin in the field of psychology (Torgerson, 1958) but has been applied to many scientific fields, including sensory science. Popper and Heymann (1996) gave an excellent review of the applications of MDS to sensory analysis. In classical applications of MDS, panelists are asked to sort a set of samples into groups according to similarity in, for example, aroma (Heymann, 1994). The MDS analysis is performed in this case on the dissimilarity scores given by a panel to pairwise presentation of samples (Popper and Heymann, 1996). Lawless and Heymann (1998) discuss issues with such pairwise comparisons. The number of pairs that can be evaluated is obviously limited, and so is the number of products that can be compared unless incomplete designs are used or otherwise derived measures of similarity or dissimilarity are used. Derived measures of similarities and MDS have been described in Lawless et al. (1995) and Heyman (1994). We focus here on the application of these methods to consumer data. In this chapter, we briefly describe the mathematical foundations of MDS and speak of a special case of MDS, unfolding. This will lay the foundation for introducing unfolding methods in a probabilistic framework (MacKay, 2001). I (Meullenet) recently had the chance to attend a workshop on probabilistic unfolding, organized by the Institute for Perception. Besides gracious hosts (Daniel Ennis and Benoit Rousseau) and good food, I also had a good introduction to probabilistic unfolding and to alternative approaches to optimizing the sensory properties of products. Later in the chapter, we give a description of these alternative approaches. As was done in previous chapters, we start with an applied explanation of the analysis concept and then move on to performing the analysis on our tortilla chip data set. We then contrast these results with those obtained in the previous chapter. MacKay (2001) argues that although many techniques are available for representing attribute and products (e.g., biplots) in a sensory space and relates these representations to hedonic data, the author does not provide an explanation of the process by which the hedonic level was derived. This approach relies mainly on the concept of a sensory ideal point that subjects use as a reference when providing hedonic responses. The probabilistic approach they employ relies on the fact that liking for a product by a consumer is not constant over time. This is because the product sensory properties can vary slightly or because the perception of the product sensory properties over a period of time may also not be constant. The advantage of probabilistic over deterministic models is their ability to account for this variability. Although similar ideas have been applied to various problems in sensory analysis, such as discrimination and preference, we constrain ourselves to applications dealing with understanding hedonic data and, in particular, with identifying product ideals, similar to methods discussed in Chapters 3 and 4.

5.2. Multidimensional Scaling (MDS) and Unfolding

MDS is a method that allows the representation of similarity or dissimilarity data among objects in a low-dimensional space (i.e., two or three dimensions; Borg and Groenen (1997). MDS encompasses a collection of methods that give insight on relations among entities by representing similarities or dissimilarities as distances in a two- or three-dimensional space (Van Deun and Delbeke, 2000). In the case of sensory data, this could be the case for p products evaluated by a group of consumers. The data are in the form of a square matrix with p columns and p rows. The measure of similarity or proximity used for modeling could be correlation coefficients between hedonic scores given to any given pair of products, or it could be the distance of a particular product from one consumer's ideal. In the case of MDS, the data matrix is $p \times p$ in dimension. The proximities are then represented in a geometrical (i.e., usually Euclidean) space, and d_{ij} represents the Euclidean distance between products i and j. The variable d_{ij} is a function of the observed proximity measures $d_{ij} = f(p_{ij})$. The coordinates of the various objects in the geometrical space and the function f, which transforms proximities into distances, are estimated by minimizing a badness of fit function, referred to as Stress in the context of MDS (Borg and Groenen, 1997):

$$S = \left[\frac{\sum_{i=1}^{n}\sum_{j>i}^{n}(\delta_{ij} - d_{ij})^2}{\sum_{i=1}^{n}\sum_{j>i}^{n} d_{ij}^2} \right], \qquad \text{(Eq. 5.1)}$$

where δ_{ij} are the optimal approximations of the transformed proximities p_{ij}. Depending on the nature of the proximities (i.e., how they were inferred), different types of MDS models are appropriate. The simplest type of MDS model is metric MDS. In this case, the data come from scaling (i.e., the similarities are quantitative) and are complete dissimilarity measures (i.e., no missing data), and the matrix is symmetric (i.e., the dissimilarity between products i and j is the same as that between products j and i; Young, 1985) The matrix is first double-centered to yield matrix $\boldsymbol{D}$ (Van Deun and Delbeke, 2000):

$$d_{ij}^2 - \frac{\sum_{i=1}^{n} d_{ij}^2}{n} - \frac{\sum_{j=1}^{n} d_{ij}^2}{n} + \frac{\sum_{i=1}^{n}\sum_{j=1}^{n} d_{ij}^2}{n^2} = -2\sum_{a=1}^{m} x_{ia} x_{ja} \qquad \text{(Eq. 5.2)}$$

The singular value decomposition of this matrix will produce the principal coordinates in a low-dimensional space. Classical metric MDS was applied to the tortilla chip data for illustration of the analysis. Because the matrix necessary to perform metric MDS is a square matrix of inner product distances, the data used were a matrix based on product correlations for overall liking. The data are presented in Table 5.1.

The entries in the data table are one minus the correlation in overall liking scores between two products. These are the data needed, as a measure of dissimilarity and not similarity is needed for classical MDS. The results of this analysis (i.e., performed using Bi-plot v1.1) are presented in Figure 5.1. The product configuration in the space is quite dissimilar to that reported in Figure 3.2 in the MDPREF analysis of the same data.

Therefore, MDS may be an alternative way to determine the spatial relationship of products in a space. The obvious limitation of MDS of an inner product dissimilarity

Table 5.1. Dissimilarity data $(1 - r_{ij})$ used to perform metric multidimensional scaling on 11 tortilla chip products.

	BYW	GMG	GUY	MED	MST	MTR	OCF	SAN	TBS	TOM	TRS
BYW	0.000	0.943	0.930	0.975	0.965	0.998	0.972	0.961	0.988	0.927	0.999
GMG	0.943[1]	0.000	0.982	0.999	0.974	0.863	0.899	1.000	0.994	0.987	0.996
GUY	0.930	0.982	0.000	0.956	0.999	0.999	0.854	0.986	0.844	0.966	0.939
MED	0.975	0.999	0.956	0.000	0.990	0.999	0.987	0.945	0.999	0.999	0.973
MST	0.965	0.974	0.999	0.990	0.000	0.982	0.999	0.995	0.968	0.928	0.946
MTR	0.998	0.863	0.999	0.999	0.982	0.000	0.998	0.997	0.988	0.961	0.997
OCF	0.972	0.899	0.854	0.987	0.999	0.998	0.000	1.000	0.982	0.935	0.981
SAN	0.961	1.000	0.986	0.945	0.995	0.997	1.000	0.000	0.944	0.966	0.966
TBS	0.988	0.994	0.844	0.999	0.968	0.988	0.982	0.944	0.000	0.996	0.948
TOM	0.927	0.987	0.966	0.999	0.928	0.961	0.935	0.966	0.996	0.000	0.911
TRS	0.999	0.996	0.939	0.973	0.946	0.997	0.981	0.966	0.948	0.911	0.000

[1] The correlation in overall liking scores across 80 panelists for a pair of products.
Note: Data table entries represent one minus the correlation in overall liking scores across 80 panelists for a pair of products.
MED, Medallion; GUY, Guy's Restaurant; OAK, Oak Creek Farms; GMG, Green Mountain Gringo; MIS, Mission strips; SAN, Santita's; TOB, Tostito's Bite Size; TOR, Tostito's Restaurant Style; BYW, Best Yet; TOM, Tom's.

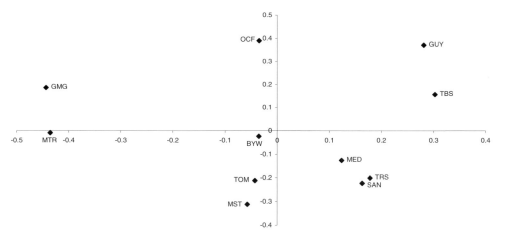

Figure 5.1. Metric multidimensional scaling of tortilla chips consumer data.

matrix is that consumer individuality is lost because only correlations across the consumer panel are being considered to build the MDS map. However, a special case of multidimensional scaling is given by unfolding models. Unfolding models are designed to estimate the coordinates in a multidimensional space of products evaluated by consumers, as well as the coordinates of the ideal product for each respondent (MacKay, 2001). The coordinates are estimated so that the distances between the ideal product and the products evaluated (i.e., real products) are inversely proportional to hedonic ratings. In the resulting maps, large distances between the ideal and real products indicate minimal liking. Unfolding models have been developed to deal with preferential or hedonic data. These models assume that hedonic responses given to a product are made in comparison to an ideal stimulus or product that is specific to each respondent. The ideal stimulus is one that would be preferred to any other product. In the unfolding framework, the hedonic scores given to the various products are a function of how far the various products are from an individual ideal. The unfolding representation seeks to represent—in a common space—the products evaluated and the consumer ideals, so that the space best depicts the original data. Unfolding is a special case of MDS because inner product distances are not available. Instead, the consumer × product matrix is the only available data, and MDS is performed on an incomplete dissimilarity matrix with $n + p$ rows and columns (n = number of consumers, p = number of products). Unfolding can be performed using most multidimensional scaling software and can be implemented in SAS (Proc MDS) and SPSS ALSCAL. However, unfolding is known to have problems with yielding degenerate solutions in which ideal points are collapsed in the middle of the multidimensional space (Young, 1985). MDPREF is a preferred methodology to avoid this problem. However, MDPREF is not an ideal methodology, as the vector model provides the general direction of a consumer's ideal but not its exact location in the space. A probabilistic approach to unfolding has been proposed and its superiority over deterministic models argued (MacKay, 2001).

5.3. Probabilistic Approach to Unfolding and Identifying the Drivers of Liking

We discuss here the use of IFPrograms—software developed by the Institute for Perception (Richmond, VA). Similar software (PROSCAL) has been developed by Professor David B. MacKay from Indiana University. In the traditional unfolding models, both real and ideal products are represented in the space as points, whereas in the probabilistic unfolding models, the real and ideal products are conceptualized as multivariate normal distributions. The main difference between the two approaches is that in the deterministic model, preference for a product by a consumer is absolute, whereas in the probabilistic approach, the assumption that product preference is always the same for a given individual is not made. Instead, the interpretation of distances between an ideal and two real products is that the real product closest to the ideal would be preferred more often than not. MacKay (2001) argues that preferential choices are not always the same across replications and that ignoring this aspect of hedonic data yields biased results. The probabilistic approach also has the advantage of establishing real product variability from a sensory perception standpoint. Often, products with large sensory variance are not preferred. However, products with no sensory variance may not appeal to more than a single segment of consumers. In some cases, increasing sensory variance (within reason) can improve its preferential choice proportion (MacKay, 2001). Let us first examine the basic concept of this type of analysis and start by explaining the concept of similarity models, which are used to perform these analyses. When a hedonic rating is given on a 9-point hedonic scale, one can often assume that a score of 9 represents the ideal. Of course, in the probabilistic framework, this ideal is an ever-moving target because the consumer's frame of mind is momentary. However, at the point in time of the evaluation, the consumer compares the product he or she tastes to an ideal (i.e., 9 on the hedonic scale). We can assume that they score the product in comparison to this imaginary ideal. The score given to the product is also considered momentary, as the score they would give to the same product an instant later could potentially differ. We attempt to describe in general terms the process employed to perform the analysis, mainly because we do not exactly know what the exact process is. To begin with the analysis, the hedonic score given to a product is transformed to a similarity to ideal score, calculated by taking the ratio of the hedonic score given to the number of categories on the scale (e.g., in this case, 9). For example, a product scoring a 9 will have a similarity to ideal of 1, whereas a product scoring a 5 would have a similarity to ideal of 0.55.

In this framework, maps considering two or more dimensions are then used to determine the product configuration in the space. The space here is a space defined by psychological dimensions and not a sensory space as previously described, as internal data (i.e., similarity scores) are used. These dimensions are psychological dimensions capable of explaining the underlying data structure. Let us use a two-dimensional space to keep things understandable. The analysis starts by randomly positioning the products in the space and assuming a variance for the products in the two dimensions. The location of the ideal is first taken as being the same as the product that received the highest hedonic score. The product coordinates and variance in the space can then be varied. At each iteration of the analysis, a distance between the ideal and the products can be calculated (i.e., either as the Euclidian distance or the city block distance). When the distance between the ideal and a product is known, a similarity value can be predicted, using the relationship given by similarity models (Ennis et al., 1988), between similarity and product distance from the ideal. An example of such a function is given in Figure 5.2. Presented here is an exponential decay function that was shown to be adequate for this purpose (Ennis et al., 1988). For example, if the distance between a product and the ideal is 1.5, the expected similarity is 0.367. For this product and ideal locations, the difference between the observed similarity and the expected similarity can be calculated. The products and ideal location and the product multivariate normal distribution in the space are modified until the sum (for all products) of squares of the differences (between the expected and observed similarities) is minimized.

This is obviously easier said than done, and we have to acknowledge here that we do not know how this is exactly implemented in Landscape Segment Analysis (LSA, IFPrograms). The configuration for which the sum of squares is minimized represents the location of the ideal for a particular subject with respect to the products represented in the test. The process is repeated for every consumer, and a landscape segment map is created. The product location on the map is the average location on each dimension across consumers, and individual ideals are represented in addition to the localized consumer density (i.e., as contours), so that consumer ideal–dense areas can be identified. Up to this point, no external data—such as descriptive analysis profiles—have been used. The next step in

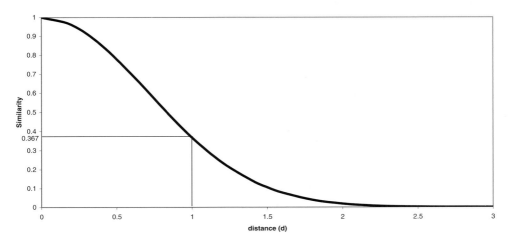

Figure 5.2. Similarity model represented as an exponential decay function.

the analysis is to project in the space the external axes (i.e., descriptive attributes or other external data). To do this, the attribute is first projected in an arbitrary direction in the space. The projection of the individual products on the axis and the sensory scores for the particular attribute are then correlated. The attribute projection is rotated until the correlation is maximized, identifying the direction of increasing intensities for the attribute. A more straightforward way to implement this is to repress the sensory attribute intensities on the product scores for the first two dimensions. The software commercialized by the Institute for Perception—IFPrograms—allows the user to have access to several types of information postanalysis. One has the ability to see the sensory profile and predicted liking score of any virtual product one may place on the map. One also has the ability to take external information such as a sensory profile for a prototype product and place it in the map. This could be especially useful for a product developer who is trying to make decisions about prototypes that have not been tested by consumers. One also has the option of selecting only a specific region of the landscape and making predictions for liking and sensory profiles only considering consumers in this area of the map. We consider these options to be unique and very useful for interpreting the meaning of the landscape, and it should be of use to product developers and market researchers to go through an infinite number of "what-if" scenarios. From that standpoint, this software and method seem superior to methods described in earlier chapters.

5.4. Examples

These examples will help make this brief and incomplete explanation of the method become clear. Our intension here is to perform LSA and compare its optimal result (i.e., the sensory profile of the optimal product) to the results obtained using the Danzart and Euclidian Distance methods discussed in Chapter 4.

5.4.1. Comparison of LSA to External Mapping Methods

The LSA was performed using IFPrograms (version 7.9). The tortilla data set was imported into IFPrograms by placing both hedonic data for individual consumers and mean descriptive analysis results in a single spreadsheet and using the data conversion tool in IFPrograms to render the data acceptable for LSA. After conversion, the software creates two separate data sets, one for the liking data and one for the sensory-of-scale data. Before LSA is launched, the user has the ability to set several options. First, a significance level is set to select the sensory attributes that are significant drivers of liking. This level is used at the end of the analysis, when the sensory attribute axes are projected in the landscape. Using the default value of 0.05 will include in the landscape only the attributes for which the correlation between the projected values on a sensory attribute axis and the actual mean scores for the products on this attribute is significantly different from zero. The attributes not included in the landscape because of poor correlations are considered to be nondrivers of liking. The other two options for the analysis concern the inclusion or exclusion of specific consumer data in the creation of the landscape. The first of these two relates to low-variance raters. Those of us with experience in consumer testing know that for any given test, a proportion of consumers will not discriminate between the products from a hedonic standpoint. In other words, these "nondiscriminators" have a tendency to either hate or love all products—or at least to score all the products the same. These nondiscriminators are not usually useful to product developers, as they are either dislikers or likers of all products. In LSA, the nondiscriminators, referred to as the low-variance raters in IFPrograms, have a tendency to have ideal points located equidistant from all products, resulting in ideals for nondiscriminators in the center of the landscape. If the number of nondiscriminators is large enough, the center of the landscape may be artificially dense, and one could conclude that an ideal for the group is in the center of the map, which would be misleading. If the option of excluding low-variance raters is chosen, the user must determine what percentage of consumers should be excluded (the default value being 5%). The second option has to do with excluding consumers who do not fit well in the space. This practice is intended to exclude those consumers for which the correlation between the observed and expected similarity measures is not significant at a user-defined level (the default being 0.05). The resulting analysis is given in Figure 5.3.

The landscape is presented in the form of a contour. These contour plots represent the localized density of consumer ideals. Consumer ideal dense areas are represented by a lighter color, whereas areas that are not ideal dense are portrayed with darker blues. The density at any given point in the landscape is determined by the size of the circle, centered on that location, that is necessary to capture a given percentage of the ideals (e.g., 5%). If the circle necessary to capture the selected percentage of ideals is small, then the area is dense, and the density value decreases as the size of the circle grows. Figure 5.3a represents such a landscape for the tortilla chip data set. One notices in the map several (up to five) distinct dense areas. This is probably not always the case and is probably here a result of the analysis being performed with only 80 consumers; the Institute for Perception recommends 300 observations (i.e., consumers) to obtain reliable maps. The densest area in the map seems to be centered at about 0.15 on dimension 1 and −0.2 on dimension 2. Figure 5.3b presents the same map with the location of the consumer individual ideals placed on the map. We find on this map that dense areas have a greater number of individual ideals in their vicinity than nondense areas. This is logical because the contours were established using the individual ideals location. Figure 5.3c is a representation of the contours superimposed on the products' locations in the landscape. The products whose position in the landscape is in a low-density area are those with low average liking scores. MED (Medallion) had an overall liking score of 5.3 whereas GUY (Guy's Restaurant), OAK (Oak Creek Farms), GMG (Green Mountain Gringo)and MIS (Mission strips) had overall liking mean scores of 5.9, 5.5, 5.5, and 6.3, respectively. In contrast, products located in the consumer-dense area of the landscape, SAN (Santita's), TOB (Tostito's Bite Size), TOR (Tostito's Restaurant Style), BYW (Best Yet), and TOM (Tom's), had average liking scores of 7.1, 7.3, 7.2, 6.4, and 6.3, respectively. Figure 5.3d shows the projection of the sensory attribute axes in the space for the attributes that were deemed to be significant drivers of liking. Among these were salt, grain complex, toasted corn, toasted grain, raw corn, and masa for flavor; crispness, fracturability, and cohesiveness of mass for texture; and degree of whiteness for visual. The angles between the first dimension and the attribute direction, as well as the correlation between the observed and projected scores of the attributes axes, are presented in Table 5.2.

The correlation coefficients ranged from 0.14 to 0.92. If one chooses to exclude the nonsignificant attributes, with $\alpha = 0.05$, the critical correlation would be 0.48 for a number of products equal to 11 (df = 9). Therefore, attributes such as amount of bubbles; sweet; micro roughness; oily, greasy lips; loose particles; roughness of mass; moisture absorption; toothpack; and oily, greasy film would not have significant correlations and would be considered nondrivers of liking. Figure 5.3e represents the product locations in the landscape in addition to a virtual product for which liking is maximized in the landscape. This is, of course, if all consumers are considered to determine the optimal product and no consumer segments are identified. The ability to find an optimal product in the landscape post-LSA

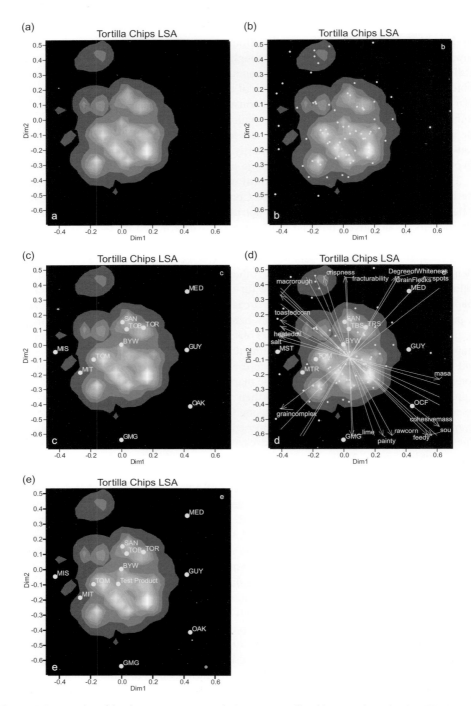

Figure 5.3. Results of landscape segment analysis on 11 tortilla chips, conducted using IFPrograms. (a) Consumer ideals density plot; (b) consumer ideals density and individual ideals; (c) consumer ideals density plot and product locations; (d) consumer ideals density, individual ideals, product locations, and sensory attribute projections; and (e) consumer ideals density, product and group ideal locations.

Table 5.2. Sensory attribute projection angles in the landscape and correlations between observed and projected scores.

Attributes	Angle (0 – 2π)	Correlation
Visual		
Degree of whiteness	1.06	0.74
Grain flecks	1.03	0.79
Char marks	4.47	0.51
Micro surface particles	2.41	0.49
Amount of bubbles	0.76	0.21
Spots	0.83	0.66
Flavor		
Sweet	5.34	0.21
Salt	2.83	0.84
Sour	5.48	0.80
Lime	4.74	0.70
Astringent	2.66	0.52
Grain complex	3.81	0.86
Toasted corn	2.63	0.92
Raw corn	5.19	0.90
Masa	6.03	0.61
Toasted grain	5.35	0.57
Painty	5.10	0.66
Feedy	5.50	0.66
Heated oil	2.71	0.60
Scorched	3.21	0.56
Cardboard	0.82	0.48
Sour grain	5.38	0.57
Texture		
Micro roughness	0.32	0.21
Macro roughness	2.39	0.62
Oily, greasy lip	2.11	0.46
Loose particles	1.61	0.43
Hardness	4.76	0.50
Crispness	1.87	0.77
Fracturability	1.62	0.82
Cohesiveness of mass	5.61	0.69
Roughness of mass	1.72	0.32
Moistness of mass	5.05	0.54
Moisture absorption	2.97	0.40
Persistence of crisp	1.63	0.47
Toothpack	0.37	0.14
Oily, greasy film	1.86	0.36

is a very useful feature. In addition, the user can select or consider only a portion of the landscape to determine the location of the optimal product as well as its sensory profile. Because the analyses performed in Chapter 3 did not consider more than one consumer segment, this feature (i.e., the consideration of a portion of the landscape) of the IFPrograms was not used here, and the whole landscape was considered to identify the sensory profile of a product that would maximize liking. This profile is given in Table 5.3 and is contrasted against the profile obtained by the Euclidian Distance method described in Chapter 4. The main result of this comparison is that the two methods yielded very similar profiles, which speaks well for the soundness of the various methods of analysis described in these chapters. The main differences between the two methods are that LSA predicted a

Table 5.3. Optimal sensory profile determined Landscape Segment Analysis in comparison to that given by the g-inverse method presented in chapter 4.

	g-Inverse	Landscape Segment Analysis	Minimum	Maximum
Visual				
Degree of whiteness	5.66	5.23	2.00	7.00
Grain flecks	6.10	5.43	3.00	8.00
Char Marks	3.17	3.68	0.00	7.00
Micro surface particles	2.16	1.96	0.50	3.50
Amount of bubbles	6.00	6.00	5.00	7.50
Spots	9.92	9.91	5.00	13.00
Visual				
Sweet	0.49	0.51	0.40	0.65
Salt	8.98	8.45	6.88	9.88
Sour	0.01	0.07	0.00	0.31
Lime	−0.01	0.01	0.00	0.22
Astringent	2.58	2.56	2.45	2.65
Grain complex	6.89	6.88	6.47	7.21
Toasted corn	3.01	2.64	1.25	3.82
Raw corn	−0.03	0.02	0.00	0.26
Masa	3.53	3.67	3.12	4.21
Toasted grain	1.44	1.56	0.75	2.34
Painty	0.01	0.07	0.00	0.45
Feedy	−0.03	0.02	0.00	0.31
Heated oil	4.45	4.41	4.15	4.61
Scorched	0.63	0.45	0.00	2.74
Cardboard	2.47	2.47	1.91	3.34
Sour grain	−0.03	0.05	0.00	0.53
Texture				
Micro roughness	8.06	8.01	6.89	8.77
Macro roughness	4.22	4.00	2.81	5.51
Oily greasy lip	6.10	5.95	4.92	6.85
Loose particles	5.76	5.68	4.43	6.71
Hardness	8.56	8.62	7.99	9.55
Crispness	10.72	10.49	9.51	11.52
Fracturability	7.78	7.66	7.13	7.98
Cohesiveness of mass	2.98	3.08	2.68	3.45
Roughness of mass	7.49	7.46	7.20	7.74
Moistness of mass	7.19	7.31	6.88	7.74
Moisture absorption	9.40	9.31	9.03	9.73
Persistence of crisp	5.47	5.41	4.94	6.00
Toothpack	5.16	5.20	4.81	5.37
Oily greasy film	4.01	3.98	3.78	4.51

lower (5.43) grain flecks intensity than Euclidian Distance Correlation Mapping (EDCM; 6.10) and a lower salt level (8.45 versus 8.98). All other attribute differences between LSA and EDCM were well below the discrimination ability of the panel. This is a rather interesting result considering that the space was created with external data (i.e., sensory data) for EDCM and with internal data (i.e., consumer liking scores) for LSA.

5.4.2. Comparison of LSA to Internal Mapping Methods

One of the deterministic mapping methods (EDCM; Meullenet et al., 2005) described in Chapter 4 has the property of being applicable in either the internal or external mapping

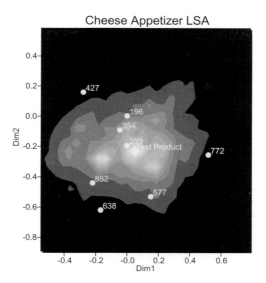

Figure 5.4. Results of landscape segment analysis on eight fried mozzarella cheese sticks, conducted using IFPrograms with product locations including the group ideal product (Test Product) and consumer ideals density contours.

framework. This is because the starting point of the analysis is a product configuration in a space that can be derived from either a sensory profiling (external) or consumer hedonic data (internal). We describe here a comparison of LSA results to those obtained from internal EDCM using the fried mozzarella appetizers described in section 1.4. This data set is better suited to performing LSA, as the number of consumers used in this study was larger ($N = 165$; the recommendation for LSA is at least 200 consumers; IFPrograms, 2005) than for the example previously described ($N = 80$). The resulting LSA map is presented in Figure 5.4. In this map, the product locations as well as the location of the product maximizing liking (i.e., the test product in Figure 5.4) are given in the first two dimensions. In addition, contour areas indicating the local density of consumer ideals are given, with the light areas representing consumer ideal–dense areas and the darker contours representing less consumer ideal–dense areas on the map. The map indicates three consumer ideal–dense areas, with the densest being located close to the center of the map. Products closest to the virtual product maximizing liking (i.e., ideal product) included products 239 (OL = 6.4; Table 1.10), 354 (OL = 6.4), and 196 (OL = 7.0), whereas products 427 (OL = 5.7), 638 (OL = 5.9), and 772 (OL = 5.9) were farthest from the ideal product.

The corresponding EDCM analysis results are given in Figure 5.5. For EDCM, the darker contour indicates consumer ideal–dense areas, whereas the lighter areas are less dense. The map also shows the product placements as well as the location of a virtual product (marked as optimum in Figure 5.5) at the most consumer ideal–dense point in the map. For EDCM, the ideal product was closest to product 239 (OL = 6.4), followed by products 354 (OL = 6.4) and 852 (OL = 6.3). The ideal product was farthest from products 772 (OL = 5.9), 638 (OL = 5.9), and 427 (OL = 5.7). Although the maps given in Figures 5.4 and 5.5 are quite different, it seems that the position of the ideal with respect to tested products is similar.

To better visualize how similar or dissimilar the two maps really are, the coordinates of the products tested in the first two dimensions were submitted to generalized procrustes

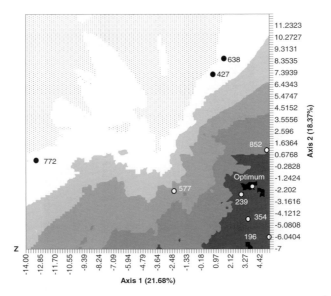

Figure 5.5. Results of Euclidian Distance Correlation Mapping for eight fried mozzarella cheese sticks conducted, using SensGear with product locations including the group ideal product (Optimum) and consumer ideals density contours.

analysis. This allowed for the product configurations to be rotated and stretched to minimize the residual variance between the two data sets. A graphical representation of the results is provided in Figure 5.6. On this graph, the closer two products with matching codes are to each other, the greater is the agreement between the two methods. Both LSA and EDCM seem to agree on the placement of products 196, 239, 354, and 638 in relation to each other. A greater disagreement was found for products 852, 577, and 772, although the configuration agreement was still acceptable. The only product for which there was real disagreement was for product 427. In LSA, 427 was located far away from product 638, whereas for EDCM, products 427 and 638 were located close to each other on the map. This difference points out the some of the drastic variance that can be obtained between deterministic and probabilistic methods. This is a point that Dr. Ennis has long argued about the superiority of probabilistic unfolding over deterministic solutions.

The fact that the product configurations in the space are somewhat different does not necessarily mean that the ultimate solution to the determination of the ideal product sensory characteristics is necessarily different. Before we discuss the solutions provided by both methods, let us first explain how the sensory profile of the optimal product are determined by LSA. Up to the point of drawing the LSA map described in Figure 5.4, no sensory information about the products in the map is used. The sensory attributes are first projected in the space. This type of projection is not uncommon and is practiced with common preference map methods such as MDPREF. Figure 5.7 provides a simple explanation of the principle. It is first assumed that a vector can be drawn in the space that represents, for a given sensory attribute, the direction of increasing intensity. The main problem at hand is finding the appropriate direction for each sensory attribute. The general idea of projections is to correlate the coordinates of the product projection on the sensory vector of interest to the actual intensities reported for the attribute. In practice, the vector

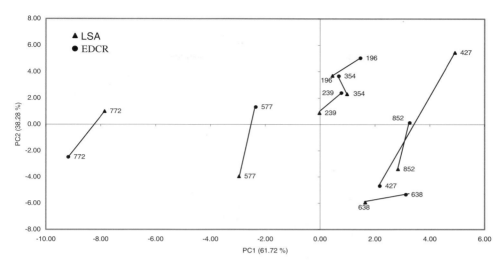

Figure 5.6. Results of generalized procrustes analysis conducted on Euclidian Distance Correlation Mapping product loadings on PC1 and PC2 and landscape segment analysis product coordinates in dimensions 1 and 2.

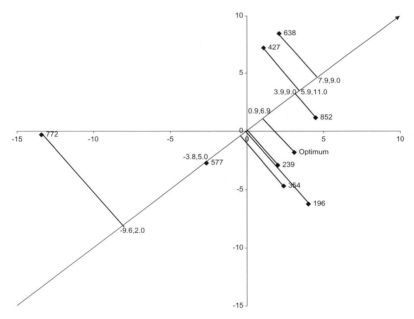

Figure 5.7. Illustration of the projection of a sensory attribute in a preference mapping space and the determination of a virtual product (Optimum) sensory intensity.

is rotated 360 degrees until the correlation between the coordinates of the projections and the actual sensory scores is maximized. This is also equivalent to regression of the sensory vector on the product coordinates in the sensory space. In Figure 5.7, the projections for each product on a sensory axis are drawn. The pair of numbers associated with each projection corresponds to the projected coordinate on that axis and the corresponding sensory score, respectively. The correlation for the eight pairs of numbers is 0.92 for an axis drawn at a 45-degree angle from the x-axis. To determine the sensory intensity of the optimal product, the coordinate of the projection on the sensory axis is calculated and the corresponding sensory intensity determined by simple linear regression.

This process is repeated for all attributes, and the sensory profile of the optimum product is determined. The maximized correlation coefficient discussed in the previous paragraph is further interpreted in methods described by Ennis and coworkers. If the correlation between projections on the sensory axis and actual sensory scores is significant (i.e., significantly different from zero at a given significance level α), the attribute is considered to be a driver of liking. If the correlation coefficient is not significant, the attribute is not considered to be an important predictor of liking—at least not in the dimensions represented by LSA. The sensory attribute projections for LSA and EDCM are presented in Figures 5.8a and 5.8b, respectively, for attributes for which the correlation coefficient was significant ($p < 0.05$).

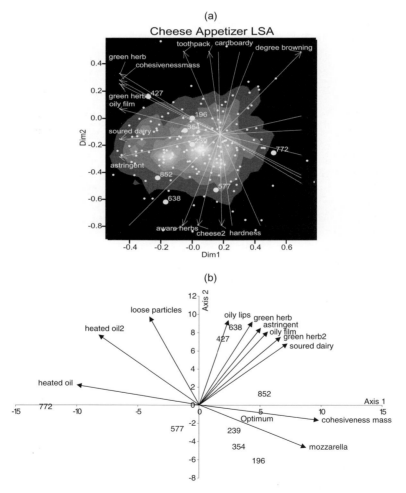

Figure 5.8. Sensory attribute projections (only significant attributes are represented) in the preference space for (a) landscape segment analysis and (b) Euclidian Distance Correlation Mapping.

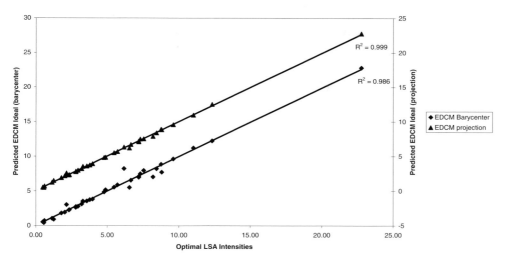

Figure 5.9. A comparison of ideal product sensory profiles given by landscape segment analysis (*x* axis) and Euclidian Distance Correlation Mapping, using either the barycenter or projection methods (*y* axes).

For LSA (Figure 5.8a), attributes degree of browning, hardness, awareness of herbs, cohesiveness of mass, oily film, toothpack, astringent, soured dairy, green herbs, cardboardy, and cheese flavor were significant drivers of liking. For EDCM (Figure 5.8b), oily lips, oily film, cohesiveness of mass, and loose particles were identified as significant texture contributors to liking. Green herb, astringent, soured dairy, mozzarella, and heated oil were significant flavor attributes. Although some attributes were common to both methods, there were a significant number of attributes for which the significance was method dependent.

Finally, the optimal sensory profiles determined through both methods were compared by plotting both the LSA optimal sensory intensities on the *x*-axis and the corresponding EDCM optimal profiles on the *y*-axis, using either the projection method or the barycenter method described previously (Figure 5.9). The coefficients of determinations for the LSA-EDCM barycenter solution and LSA-EDCM projection solution were 0.986 and 0.999, respectively. For the EDCM barycenter solution, there were five attributes for which the solution did not agree with that given by LSA. These attributes were shape integrity (EDCM = 7.0 vs. LSA = 8.2), awareness of herbs (EDCM = 8.2 vs. LSA = 6.2), degree of browning (EDCM = 7.7 vs. LSA = 8.8), visual roughness (EDCM = 5.5 vs. LSA = 6.5), and moisture release (EDCM = 3.0 vs. LSA = 2.1). The agreement between LSA and EDCM projection solutions was much greater, with the largest differences observed for shape integrity (EDCM = 7.9 vs. LSA = 8.2), visual roughness (EDCM = 6.2 vs. LSA = 6.5), and moisture release (EDCM = 2.6 vs. LSA = 2.1). Overall, we conclude that in this case, LSA provided a similar solution to a deterministic internal mapping technique.

References

Borg, I. and P. Groenen. 1997. Modern multidimensional scaling. Theory and application. New York: Springer.
Ennis, D.M., J. Palen, and K. Mullen. 1988. A multidimensional stochastic theory of similarity. J. Math. Psychol. 37:104–111.

Heymann, H. 1994. A comparison of free choice profiling and multidimensional scaling of vanilla samples. J. Sensory Stud. 9:445–453.

IFPrograms. 2005. Software manual version 7.10.

Lawless, H.T. and H. Heymann. 1998. Sensory evaluation of food, principles and practices. Gaithersburg, MD: Aspen.

Lawless, H.T., N. Sheng, and S.S.C.P. Knoops. 1995. Multidimensional scaling of sorting data applied to cheese perception. Food Q. Pref. 6:91–98.

MacKay, D.B. 2001. Probabilistic unfolding models for sensory data. Food Q. Pref. 12:427–436.

Meullenet, J.-F., R. Xiong, and T. Alpers. 2005. A mapping method for determining optimum sensory profiles. Sixth Pangborn Sensory Science Symposium (Harrogate, UK).

Popper, R. and Heymann, H. 1996. Analyzing differences among products and panelists by multidimensional scaling. In T. Naes and E. Risvik (eds.). Multivariate analysis of data in sensory science. Amsterdam: Elsevier Science, pp. 159–184.

Torgerson, W. 1958. Theory and methods of scaling. New York: Wiley.

Van Deun, K. and L. Delbeke. 2000. Multidimensional scaling. Available at http://www.mathpsyc.uni-bonn.de/doc/delbeke/delbeke.htm. Accessed January 15, 2006.

Young, F.W. 1985. Multidimensional scaling. In S. Kotz, N.L. Johnson, and C.B. Reads (eds.). Encyclopedia of statistical sciences, Volume 5. New York: Wiley, pp. 649–659.

6 Consumer Segmentation Techniques

6.1. Introduction

The desire to segment consumers into homogeneous groups of individuals comes from the fact that consumers' hedonic responses show a fair amount of interindividual differences. Simply put, we all like different things, and segmentation is a nice way to summarize our individual differences. Consumer segmentation is especially important in the preference mapping framework because one of the goals of these methods is usually to identify sensory attributes driving liking. Market researchers use segmentation to model differences in consumers and often use this information to develop products targeted toward specific population segments (Moskowitz and Bernstein, 2000). In this chapter, we describe several methods of segmentation, including the well-known methods of hierarchical cluster analysis and the lesser-known latent class (LC) models. As in previous chapters in this book, the methods will first be described and then applied and compared through their application to sample data sets.

6.2. Methods Available

6.2.1. Cluster analysis

Cluster analysis seeks to identify homogeneous subgroups of consumers in a population. Cluster analysis identifies groups that both minimize within-group variation and maximize between-group variation.

In agglomerative hierarchical clustering, every consumer is initially considered to be in a separate cluster. The two consumers with the smallest distance between them are grouped into a cluster. The consumer the smallest distance from either of the first two consumers is considered next. If the data for that individual are closer to that for a fourth consumer than they are to either of the first two, the third and fourth consumers are grouped into a second cluster. If not, the third consumer is grouped in the first cluster. The process is then repeated, adding consumers to existing clusters or creating new clusters until every consumer has been considered.

The first step in cluster analysis is the establishment of the similarity or distance matrix. This matrix is a table in which both the rows and columns are the units of analysis and the cell entries are a measure of similarity or distance for any pair of consumers. Euclidean distances are the most commonly used distance measures. Let us assume that our data are composed of rows that are consumers and columns that are stimuli (i.e., products) hedonic scores. The Euclidian distance between two consumers is the square root of the sum of the squared differences for each variable. Squared Euclidean distances are most often used to create the dissimilarity matrix. When two or more variables are used (in our case, hedonic scores for products are the variables) to define the distance, the variable with the larger magnitude will dominate. This problem can be solved with variable standardization. However, with clustering of consumer liking data, this standardization is not commonly done, as the magnitude of the differences in liking of a product by two consumers is important.

There are a variety of different measures of interobservation distances and intercluster distances to use as criteria when merging nearest clusters into broader groups or when considering the relation of a point to a cluster. Examples of this are the nearest or furthest neighbors, centroid method, and Ward's criteria. Ward's criteria, which is the most often used method found in the literature for segmenting consumers, calculates the sum of squared distances from each consumer in a cluster to the mean of all variables. The cluster that is merged is the one that will increase the sum the least. Ward (1963) proposed a clustering procedure seeking to form partitions (clusters) P_k, $P_{k-1}, \ldots, P_1$ in a manner that minimizes the loss of information associated with each grouping. Information loss is defined by Ward in terms of an error sum-of-squares criterion. Error sum of squares is defined as follows:

$$ESS = \sum_{k=1}^{K} \sum_{x_i \in C_k} \sum_{j=1}^{p} (x_{ij} - \overline{x}_{kj})^2, \qquad (Eq.\ 6.1)$$

where the cluster mean is

$$\overline{x}_{kj} = \frac{1}{n_k} \sum x_i \in C_k^{x_{ij}}, \qquad (Eq.\ 6.2)$$

where x_{ij} denotes the value for the ith individual in the jth cluster, k is the total number of clusters at each stage, and n_j is the number of individuals in the jth cluster.

6.2.2. Latent Class Models

LC models are finite mixture models that have been applied to cluster analysis and other multivariate analysis problems such as factor and regression analyses. Latent constructs are created from indicator variables, as in structural equation modeling, and then used to assign observations (i.e., consumers) to LCs, which are often interpreted as clusters or segments of consumers. Membership of each individual to one of the K-category latent variables is determined by the membership probability of a consumer to each LC. Latent profile models are a variant of LC models for continuous variables, which will later be explored as an alternative to LC models. LC modelling software such as Latent Gold (Statistical Innovations, Belmond, MA) implements LC models for cluster analysis, factor analysis, and regression with LCs.

6.2.2.1. LC Cluster Analysis

The LC cluster module in Latent Gold differs from traditional cluster analysis algorithms, which group cases near each other by some measure of dissimilarity. The LC approach defines one cluster per LC, using model-based probabilities to classify consumers. Covariates such as consumer demographic information can also be used in the model. Because Latent Gold was the commercial software used to perform the analyses for the examples to follow, we give here some details about its use. In Latent Gold, the Variables tab is used to specify the number of classes for cluster models. One may also specify a range to estimate several models (e.g., enter "1–4" to estimate four LC models containing 1, 2, 3, and 4 clusters, respectively). This can be useful to determine which number of clusters best describes the data. Options exist in Latent Gold to constrain the model in various ways. For cluster models, the Clusters tab of the Analysis dialog box allows the user to change the number of clusters or to restrict various effects to equal zero (beta effects of the indicators on the latent variables; gamma effects of selected covariates on the latent variable).

6.2.2.2. LC Factor Analysis

The LC factor model identifies factors that group variables that share a common source of variation (Vermunt and Magidson, 2000). Several factors are usually included in the model when dealing with segmentation of consumer data. This is done so that the first factor can account for differences in hedonic response levels between consumers, whereas the subsequent factors segment consumers more by preference patterns. In the case of several factors, the LC factor model provides membership classification probabilities for each consumer directly into each level of each factor. In Latent Gold, LC factor analysis treats the LCs as factors but avoids such problems of traditional factor analysis as having to rotate factors and having to assume continuous observed variables. One may specify the number of factors for factor models, where each factor may have up to five levels. Correlations between selected pairs of factors can be included in the factor model. Output may include biplots, which can locate values of nominal, ordinal, or continuous variables on factor axes.

Latent Gold allows the users to change the number of factors or factor levels, restrict factor effects to zero on selected variables, and restrict selected factor correlations to zero. In addition, correlations between selected pairs of factors can be included in the factor model. In addition, Latent Gold 4.0 has options for controlling the number of iterations and convergence limits, starting values, Bayes constants, treatment of missing data, and error variances.

6.3. Segmentation Methods Using Hierarchical Cluster Analysis

The vast majority of papers dealing with segmentation of consumers based on preference patterns employs hierarchical cluster analysis (Helgesen et al., 1997; Jaeger et al., 2003; Martinez et al., 2002). There are several considerations associated with results from such clustering methods. According to Westad et al. (2004), it is vital to determine whether the segments identified by cluster analysis are valid or simply a result of data overfitting. One way to test the relevance of a segmentation scheme is to determine whether the hedonic scores given to the products within a segment are significantly different from each other. Alternatively, relevance can be established if the segments are differentiated by demographic variables (e.g., age, gender, or consumption habits).

There are many different approaches to assessing segments, and we can only discuss a few. Variations in approaches start first with the type of data employed to do the analysis. Often, the preference data (the matrix with consumers as rows and products as columns) are used directly. In some cases, either product scores or variable loadings from a principal component analysis of the consumer × product (long, skinny matrix) or product × consumer (short, fat matrix) matrices, respectively, are used for clustering. In the case where the preference data are used directly, the data are of the short, fat type, and hierarchical cluster analysis can be performed on raw, centered (i.e., row mean is zero) or scaled data (i.e., row mean is zero and row standard deviation is one). The use of raw data implies that clustering will be based on the magnitude of the preference scores (i.e., segmenting consumers with high liking scores and consumers with low liking scores in different clusters). McEwan (1998) recommends this practice as a part of a phased approach to clustering to identify the proportion of consumers who have high or low liking for the products presented. The use of centered data implies that differences in the use of the hedonic scale are eliminated. Rather, clustering will be primarily influenced by the relative liking differences and rank order of preference and not by the magnitude of the raw scores. The use of standardized data for clustering was not recommended by McEwan (1998) as a viable option, mostly because the size of differences in acceptability of different product is an important aspect of the data when a verbal type of hedonic scale is used. However, McEwan argues that clustering performed on standardized data could be useful in cases in which unstructured scales are used. We believe that if standardization is valid on unstructured scales, the same could probably applied to labeled scales, such as the 9-point hedonic scale, where the distance between categories may not have the same meaning for all consumers. The number of clusters selected is a rather subjective process and is also a matter of debate. McEwan (1998) recommended not including cluster solutions where the smaller cluster represents less than 20% of the total number of consumers. This seems reasonable from an interpretability standpoint, as with clusters of very small size, one could question the reliability of the interpretation derived.

Let us now take an example and work through the various options available. We will work with the muscadine grape juice data set presented in Chapter 1. Because the number of consumers polled was small, we limited the number of possible clusters to three and performed a hierarchical cluster analysis with Ward's method, using Euclidian distances, as suggested by McEwan et al. (1998). First, we present the partial data matrices used for cluster analysis of raw (Table 6.1), centered (Table 6.2,) and standardized (Table 6.3) preference data.

The purpose of these tables is to briefly explain the concept of mean centering and standardization. Table 6.1 comprises the raw sensory data, with products in columns and consumers in rows. The data points in each cell are an integer number between 1 and 9 that corresponds to the equivalent labeled category on the 9-point hedonic scale, with dislike extremely being a score of 1 and like extremely being a score of 9. In this case, consumers have different mean preference and variation around the mean (i.e., standard deviation). In the case of centered data (Table 6.2), the mean preference for each consumer (Table 6.1) is simply subtracted from every observation corresponding to a particular consumer. The result is that the mean of the centered data for each consumer becomes zero. However, the standard deviations remain unchanged from the raw data. In the case of standardized data (Table 6.3), each observation of the mean centered data is divided by the corresponding standard deviation (i.e., consumer or row specific). This yields a mean of zero and a standard deviation of one for each row. The data actually used by clustering algorithms are not actually the data imputed as raw, centered, or standardized but, instead, a measure of similarity or dissimilarity between two observations (i.e., consumers). This measure of dissimilarity in our case is a Euclidian distance between two consumers, with each product representing a separate dimension. This is important to consider because these Euclidian distances are influenced by either mean centering or standardization. Let us consider a simple example in which we have three consumers evaluating two products (Figure 6.1).

We chose two products because we can then represent the data in a plane, with the scores on each axis corresponding to the scores given to the two products. From visual inspection of the raw data, it is clear that consumers 2 and 3 are slightly closer to each other than they are to consumer 1. This stems from the fact that consumers 2 and 3 scored product 1 high (giving it scores of 8 and 9, respectively), whereas consumer 1 scored product 1 extremely low (giving it a score of 2). Therefore, cluster analysis would result in

Table 6.1. Partial raw preference data used for hierarchical cluster analysis.

Consumer	Carlos	Supreme	Nestitt	Post White	Summit	Post Red	Ison	Southern Home	Granny Val	Black Beauty	Mean	SD
1	4	3	4	6	6	1	2	3	4	6	3.9	1.7
2	8	8	7	7	8	8	7	9	8	6	7.6	0.8
3	8	4	8	4	8	7	7	8	7	6	6.7	1.6
4	5	4	3	2	4	7	6	3	7	6	4.7	1.8
5	7	7	8	7	7	6	7	7	7	8	7.1	0.6
6	5	7	5	6	6	7	6	7	8	8	6.5	1.1
7	5	6	7	7	7	8	8	8	5	4	6.5	1.4
8	6	3	5	4	7	4	7	8	7	7	5.8	1.7
9	6	9	7	3	8	8	7	8	7	7	7	1.6
10	8	7	8	7	8	7	7	7	7	7	7.3	0.5
11	6	6	5	6	4	8	4	7	4	2	5.2	1.8
12	5	1	1	6	7	1	2	3	9	6	4.1	2.9
13	7	7	8	8	8	8	9	8	7	8	7.8	0.6
14	4	7	6	4	8	6	6	4	5	8	5.8	1.5
15	6	8	7	6	8	5	7	8	7	7	6.9	1.0
16	8	9	9	9	9	9	8	7	7	9	8.4	0.8
17	9	7	8	7	9	6	6	8	8	9	7.7	1.2
18	7	9	7	9	8	9	8	9	7	9	8.2	0.9
19	6	8	8	7	7	7	8	8	6	6	7.1	0.9
20	3	8	8	5	4	8	7	7	7	6	6.3	1.8

Table 6.2. Partial centered preference data used for hierarchical cluster analysis.

Consumer	Carlos	Supreme	Nestitt	Post White	Summit	Post Red	Ison	Southern Home	Granny Val	Black Beauty	Mean	SD
1	0.1	-0.9	0.1	2.1	2.1	-2.9	-1.9	-0.9	0.1	2.1	0	1.7
2	0.4	0.4	-0.6	-0.6	0.4	0.4	-0.6	1.4	0.4	-1.6	0	0.8
3	1.3	-2.7	1.3	-2.7	1.3	0.3	0.3	1.3	0.3	-0.7	0	1.6
4	0.3	-0.7	-1.7	-2.7	-0.7	2.3	1.3	-1.7	2.3	1.3	0	1.8
5	-0.1	-0.1	0.9	-0.1	-0.1	-1.1	-0.1	-0.1	-0.1	0.9	0	0.6
6	-1.5	0.5	-1.5	-0.5	-0.5	0.5	-0.5	0.5	1.5	1.5	0	1.1
7	-1.5	-0.5	0.5	0.5	0.5	1.5	1.5	1.5	-1.5	-2.5	0	1.4
8	0.2	-2.8	-0.8	-1.8	1.2	-1.8	1.2	2.2	1.2	1.2	0	1.7
9	-1	2	0	-4	1	1	0	1	0	0	0	1.6
10	0.7	-0.3	0.7	-0.3	0.7	-0.3	-0.3	-0.3	-0.3	-0.3	0	0.5
11	0.8	0.8	-0.2	0.8	-1.2	2.8	-1.2	1.8	-1.2	-3.2	0	1.8
12	0.9	-3.1	-3.1	1.9	2.9	-3.1	-2.1	-1.1	4.9	1.9	0	2.9
13	-0.8	-0.8	0.2	0.2	0.2	0.2	1.2	0.2	-0.8	0.2	0	0.6
14	-1.8	1.2	0.2	-1.8	2.2	0.2	0.2	-1.8	-0.8	2.2	0	1.5
15	-0.9	1.1	0.1	-0.9	1.1	-1.9	0.1	1.1	0.1	0.1	0	1.0
16	-0.4	0.6	0.6	0.6	0.6	0.6	-0.4	-1.4	-1.4	0.6	0	0.8
17	1.3	-0.7	0.3	-0.7	1.3	-1.7	-1.7	0.3	0.3	1.3	0	1.2
18	-1.2	0.8	-1.2	0.8	-0.2	0.8	-0.2	0.8	-1.2	0.8	0	0.9
19	-1.1	0.9	0.9	-0.1	-0.1	-0.1	0.9	0.9	-1.1	-1.1	0	0.9
20	-3.3	1.7	1.7	-1.3	-2.3	1.7	0.7	0.7	0.7	-0.3	0	1.8

Table 6.3. Partial standardized preference data used for hierarchical cluster analysis.

Consumer	Carlos	Supreme	Nestitt	Post White	Summit	Post Red	Ison	Southern Home	Granny Val	Black Beauty	Mean	SD
1	0.1	-0.5	0.1	1.2	1.2	-1.7	-1.1	-0.5	0.1	1.2	0	1.0
2	0.5	0.5	-0.7	-0.7	0.5	0.5	-0.7	1.7	0.5	-1.9	0	1.0
3	0.8	-1.7	0.8	-1.7	0.8	0.2	0.2	0.8	0.2	-0.4	0	1.0
4	0.2	-0.4	-1.0	-1.5	-0.4	1.3	0.7	-1.0	1.3	0.7	0	1.0
5	-0.2	-0.2	1.6	-0.2	-0.2	-1.9	-0.2	-0.2	-0.2	1.6	0	1.0
6	-1.4	0.5	-1.4	-0.5	-0.5	0.5	-0.5	0.5	1.4	1.4	0	1.0
7	-1.0	-0.3	0.3	0.3	0.3	1.0	1.0	1.0	-1.0	-1.7	0	1.0
8	0.1	-1.7	-0.5	-1.1	0.7	-1.1	0.7	1.3	0.7	0.7	0	1.0
9	-0.6	1.2	0.0	-2.4	0.6	0.6	0.0	0.6	0.0	0.0	0	1.0
10	1.4	-0.6	1.4	-0.6	1.4	-0.6	-0.6	-0.6	-0.6	-0.6	0	1.0
11	0.5	0.5	-0.1	0.5	-0.7	1.6	-0.7	1.0	-0.7	-1.8	0	1.0
12	0.3	-1.1	-1.1	0.7	1.0	-1.1	1.9	-0.4	1.7	0.7	0	1.0
13	-1.3	-1.3	0.3	0.3	0.3	0.3	0.1	0.3	-1.3	0.3	0	1.0
14	-1.2	0.8	0.1	-1.2	1.4	0.1	0.1	-1.2	-0.5	1.4	0	1.0
15	-0.9	1.1	0.1	-0.9	1.1	-1.9	0.1	1.1	0.1	0.1	0	1.0
16	-0.5	0.7	0.7	0.7	0.7	0.7	-0.5	-1.7	-1.7	0.7	0	1.0
17	1.1	-0.6	0.3	-0.6	1.1	-1.5	-1.5	0.3	0.3	1.1	0	1.0
18	-1.3	0.9	-1.3	0.9	-0.2	0.9	-0.2	0.9	-1.3	0.9	0	1.0
19	-1.3	1.0	1.0	-0.1	-0.1	-0.1	1.0	1.0	-1.3	-1.3	0	1.0
20	-1.9	1.0	1.0	-0.7	-1.3	1.0	0.4	0.4	0.4	-0.2	0	1.0

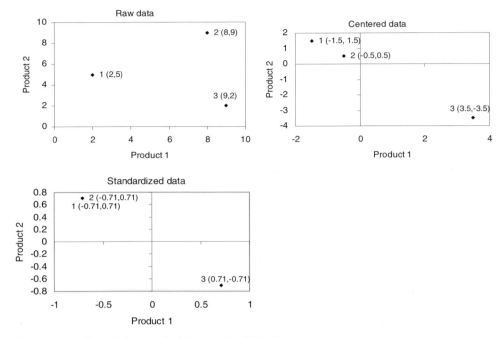

Figure 6.1. Effect of data manipulations on Euclidian distances.

first grouping of consumers 2 and 3. Let us now contrast these results with those obtained for centered data (Figure 6.1). In this case, data centering has the effect of bringing consumers 1 and 2 closer together. This is because these two consumers actually agreed that product 2 was preferable to product 1, whereas consumer 3 felt the opposite. Cluster analysis would now result in the clustering of consumers 1 and 2 together. After centering, the magnitude of the preference scores is ignored and emphasis is placed on rank order and the magnitude of the differences among products. Although data centering helps in removing scaling effects, the somewhat extreme example given here raises questions about the appropriateness of this practice. The positive aspect of mean centering is that some consumers possibly may inherently give lower preference scores to products because the consumers have higher expectations for those products or do not feel as passionate for the products as do others. The negative aspect of the practice is that clustering results in the grouping of consumers who like some products and dislike others with consumers who dislike all products but dislike some less than others. The preference of the author is for not centering the data and for letting low-scoring consumers be grouped together. However, some researchers make compelling arguments for data centering. Standardization of the data is a logical evolution of centering, where only the rank order of the products is now taken into account. In our example, this results in the overlapping of consumers 1 and 2 because they agreed on the rank order of the products. If one feels that mean centering is an adequate practice, then standardization would seem to be an acceptable practice. However, McEwan, in her review of cluster analysis and preference mapping (1998), concludes that centering is acceptable, whereas standardization should not be considered—at least not on a routine basis. However, the argument made by McEwan, that "The size of differences in acceptability between products is of importance, particularly as a labeled hedonic scale had been used, as opposed to an unstructured scale" (McEwan, 1998,

p. 38), is not particularly convincing. I argue that the magnitude of the acceptability scores removed by centering is as least as—if not more—important than the size of the differences. However, McEwan discusses another important point in consumer segmentation and proposes a sensible solution to this problem. The issue relates to nondiscriminators—consumers who do not show a clear preference for any of the products. McEwan recommends a phased approach in which cluster analysis is first performed on raw data. ANOVA is then performed within each cluster, and clusters showing no significant differences among products are eliminated from the data. As a second stage, cluster analysis is performed on centered data (i.e., consumers that remain after the first stage). The need to remove consumers that do not discriminate among products is obvious. An alternative to the approach proposed by McEwan would be to assess the variance across products for each consumer and eliminate those consumers who displayed low variance. This would eliminate consumers being assessed on an individual rather than group (i.e., cluster) basis. This strategy is indeed implemented in Landscape Segment Analysis (IFPrograms, see Chapter 5), whereby a percentage of consumers are eliminated from the analysis on the basis of individual consumer variance in preference scores.

Let us now return to our muscadine grape juice example. Hierarchical cluster analysis was performed using Euclidian distances on raw, centered, or standardized data. The number of consumers in each of three clusters is given in Table 6.4. Cluster analysis outputs are usually dendrographs, such as that depicted in Figure 6.2. From this graph, any number of clusters can be obtained by drawing a horizontal line somewhere on the y-axis. For example, a horizontal line crossing the y-axis at 400 would yield three clusters, whereas a line crossing at 1000 would yield only two clusters. The optimal number of clusters is somewhat arbitrary and was chosen for our example to be three.

Cluster analysis performed on raw data yielded clusters with 6, 33, and 22 consumers (Figure 6.3). Clusters with small membership (e.g., 6) are probably not appropriate to interpret, especially if they represent less than 20% of the total number of consumers. Cluster 1 was represented by consumers who liked Summit, Granny Val, and Black Beauty and disliked Post Red most. Clusters 2 and 3 were mostly differentiated by liking level, with cluster 2 scoring higher than cluster 3. Within clusters, analysis of variance revealed that there were some significant differences in liking among products (Table 6.4). From a product development standpoint, the usefulness of cluster analysis performed on raw data could result in concentrating development efforts to satisfy cluster 2 because they seem to be most enthusiastic about this product category. This would result in product Carlos being identified as the best product. However, one could argue that potential consumers of this product category are in cluster 2 as well. After all, consumers participating in this test were all supposed to already be product users. If clusters 2 and 3 are considered, then products such as Ison and Southern Home would be better choices, as they were among the most liked by both groups.

Results of cluster analysis on centered data (Figure 6.3) yielded one large cluster (cluster 2, $N = 49$) and two very small clusters ($N = 7$ and $N = 5$). For cluster 2, Ison and Southern Home were the most liked products. Not surprisingly, based on previous discussions of the effect of centering, clusters 2 and 3 from raw data cluster analysis were grouped in cluster 2 of the centered cluster analysis.

Results of cluster analysis on standardized data, where only the preference rank order is considered, yielded quite different results. First, the three clusters were of roughly equal sizes ($N = 23$, 16, and 22). Second, the clusters showed some clearer differences in liking among products. Cluster 1 preferred Black Beauty, Carlos, Granny Val, and Summit and most disliked Post Red and Summit. Cluster 2 liked Post Red and Southern Home most

Table 6.4. Number of consumers and product mean separation within each cluster for hierarchical cluster analysis performed on raw, centered, and standardized data.

	Cluster	N	Black Beauty	Carlos	Granny Val	Ison	Nestitt	Post Red	Post White	Southern Home	Summit	Supreme	LSD values
Raw	1	6	7.5a	5.0b	7.8a	4.8bc	3.0cd	1.3d	5.0b	4.3bc	7.3a	3.2bcd	1.94
	2	33	7.5ab	7.9a	7.5ab	7.7a	7.8a	7.1b	7.4ab	7.7a	7.7a	7.7a	0.50
	3	22	5.9a	5.5ab	6.1a	6.4a	5.9a	6.4a	4.9b	6.4a	5.8ab	5.8ab	0.98
Centered	1	7	8.0a	7.6ab	6.6b	5.1cd	6.6b	3.9d	6.7ab	6.4bc	6.6b	6.4bc	1.29
	2	49	6.7cd	6.8bcd	7.0abc	7.3a	7.0abc	7.1abc	6.4d	7.2ab	6.9abc	7.0abc	0.51
	3	5	7.8a	5.2bc	8.6a	5.4b	2.8de	1.4e	4.8bcd	4.6bcd	7.6a	3.2cde	2.14
Standardized	1	23	7.6a	7.2ab	7.5a	6.1c	6.2c	4.5d	6.4bc	6.5bc	7.2ab	5.8c	0.86
	2	16	6.0c	6.3c	6.4c	6.8bc	6.2c	7.6ab	6.9bc	7.8a	6.9bc	6.6c	0.92
	3	22	6.9bc	6.7bc	7.0bc	7.9a	7.4ab	7.2bc	5.7d	6.7c	6.7bc	7.4ab	0.72

Note: Means within a row with different letters are significantly different at $p < 0.05$.

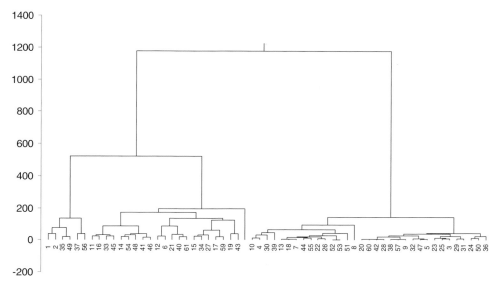

Figure 6.2. Dendrogram given from raw data from Ward's hierarchical cluster analysis, using Euclidian distances.

and disliked most Black Beauty, Carlos, and Granny Val; in other words, clusters 1 and 2 displayed opposite liking patterns. Cluster 3 differentiated itself from clusters 1 and 2 by liking Ison most, followed by Nestitt and Supreme, and disliking Post White the most. In this particular example, the use of standardized data seemed to differentiate clearer clusters of consumers with marked differences in preference patterns. Although it is difficult to gauge what the right approach might be because no data are available on product success in the marketplace, standardizing data before hierarchical cluster analysis may be an interesting avenue to explore. We revisit this question after discussing LC cluster analysis to determine which approach in hierarchical clustering is closest to the solutions given by LC models.

In the recent past, LC models have gained popularity as an alternative to hierarchical clustering. Popper et al. (2004) describe the first application of LC models to food product development. The authors used various strategies to segment consumer liking for crackers. The models considered included a LC cluster, a LC factor model, and several regression models, including a LC continuous factors model. The LC cluster model is the traditional LC model, which does not distinguish between variations caused by overall liking level and relative differences in liking levels for different products (Popper et al., 2004). This is a similar situation to that encountered when hierarchical clustering is performed on raw data. The LC factor model is designed to take into account differences in liking levels and removing its effect during segmentation. The first D factor (where D stands for discrete) is used to model liking level, whereas subsequent D factors are used to identify segments. The various regression models investigated by Popper et al. (2004) will not be discussed here. Briefly, they used a continuous intercept to model differences in liking level and used either LCs or continuous factors and either a product variable or a set of sensory descriptors as predictors. The idea of combining segmentation with the identification of sensory attributes driving liking is very interesting and is reminiscent of the preference-mapping techniques discussed in previous chapters.

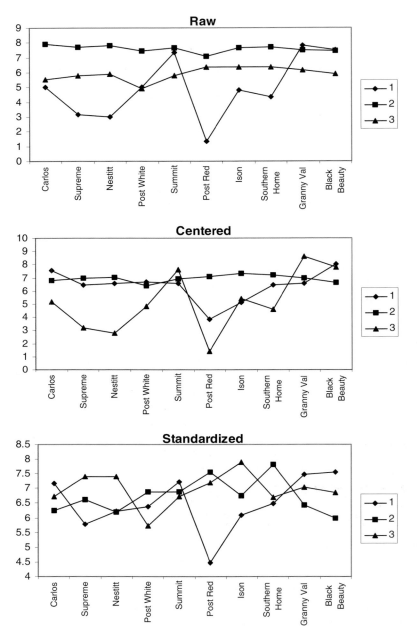

Figure 6.3. Product means by clusters for Ward's hierarchal cluster analysis performed on raw, centered, or standardized data.

Let us now apply some of these methods to our muscadine grape juice study. This discussion starts with the disclaimer that our data set was not quite large enough and did not provide enough degrees of freedom to quite satisfy the requirements of the LC models. The reader is therefore cautioned against making too much of the following discussion but, rather, to take it as an illustration of the methods. All the analyses reported here were conducted with Latent Gold 4.0, and the model fit was assessed using the Bayesian Information Criterion, referred to thereafter as BIC. The data matrix used for the LC cluster and factor models was identical to that given in Table 6.1, and no centering or standardization was performed. On the basis of the BIC, the optimal LC cluster model would have two clusters. However, the model with three clusters was not much higher in BIC (i.e., BIC = 2340 for three clusters and 2338 for two clusters). The three-cluster model was chosen so that a comparison with results obtained by hierarchical cluster analysis could be undertaken. Clusters 1, 2, and 3 contained 31, 21, and 5 consumers, respectively, which is roughly the same as those found by hierarchical cluster analysis of raw data. Product means by LC clusters are presented in Figure 6.4. The graph presented appears to be very similar to that presented for raw data hierarchical cluster analysis, except that cluster 1 corresponds to LC cluster 3, cluster 2 is like LC cluster 1, and cluster 3 is like LC cluster 2. This was verified by the fact that the correlations of mean liking scores between corresponding clusters (i.e., comparing hierarchical to LC) ranged from 0.96 to 0.99. As a consequence, it is obvious that LC clustering results in the same interpretation as that given by hierarchical cluster analysis performed on raw data. It is clear that LC clustering has the tendency to group consumers according to liking levels across products, which is not often very meaningful in establishing liking patterns and differentiating products from one another.

The next analysis performed was an LC factor model as described by Popper et al. (2004). The model chosen had two D factors, each with three levels. The final form of the model was obtained by trial and error, varying the number of factors and the number of levels in each factor. The model chosen resulted in a BIC of 2303, which was smaller that reported for the LC cluster model. As previously mentioned, the LC factor models have the advantage of factoring out scaling effects in the data. This is usually captured by the first D factor and is well illustrated in Figure 6.4. The mean values across products for the three levels of D factor 1 are clearly differentiated; consumers represented by level 1 score lower than those in level 2. The same is true for consumers represented by levels 2 and 3. However, there seem to be little value in examining factor 1 too carefully because there do not seem to be great differences among products within the clusters. The practice with LC factor models is to instead examine clusters resulting from subsequent factors. Cluster means for D factor 2 are presented as the last graph in Figure 6.4. Clusters (levels) 1, 2, and 3 comprised 21, 32, and 8 consumers, respectively. D factor 2 cluster 1 liked Post Red and Supreme most and least liked Black Beauty. In contrast, D factor 2 cluster 2 liked Granny Val and Black Beauty most and disliked Post Red and Supreme. D factor 2 cluster 3 showed similar preference patterns to D factor 2 cluster 2 (i.e., there was a mean score correlation of 0.83). Hierarchical cluster (i.e., centered and standardized data) means were correlated to D factor 2 cluster means to determine whether the interpretation of clusters was similar between the two approaches. D factor 2 cluster 1 was significantly correlated to centered ($r = 0.65$) and standardized ($r = 0.54$) cluster 2, but the correlations were not high enough to claim equivalence. D factor 2 cluster 2 means scores were most highly correlated to those of centered cluster 3 ($r = 0.84$), whereas mean scores for D factor 2 cluster 3 were most highly correlated to standardized cluster 3 ($r = 0.98$). We conclude from these correlations that the segmentation solution provided by the LC factor model was unique and different from those provided by the various forms of hierarchical cluster analysis tested. This aspect of LC modeling is interesting because it provides a new avenue for consumer segmentation. However, it is yet to be demonstrated that conclusions drawn from LC models would result in more successful products in the marketplace, compared to those products developed on the basis of results from hierarchical clustering.

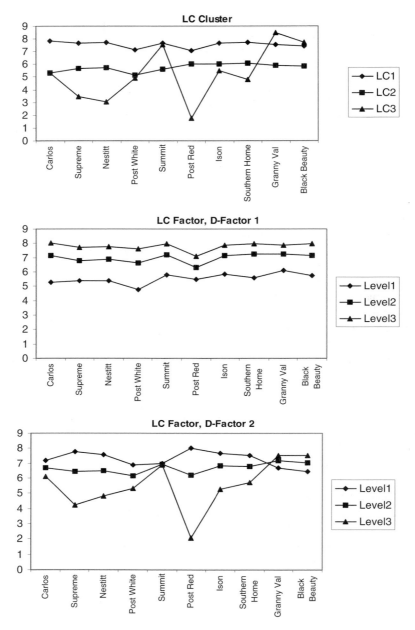

Figure 6.4. Product means by clusters for latent class cluster and factor models.

Although the methods described previously are most commonly used, many other approaches have been described and merit mention here. We apologize in advance to those not discussed here.

Moskowitz and Bernstein (2000) describe a segmentation method based on external data. The underlying principle of the segmentation is based on the inverted U-shaped function long advocated by Moskowitz. This function implies that for any given sensory modality, there exits a level at which liking is ideal, and above and below which, liking for a product tends to decrease. Moskowitz says that modeling can be undertaken at the individual level and that an optimal sensory intensity for an attribute can be derived for each consumer. This process is repeated for all attributes and the optima are identified on each attribute and used as the basis for a hierarchical cluster analysis. The analysis includes a step of data reduction via a principal component analysis, with consumers as rows and sensory attribute optima as variables (Moskowitz et al., 1985). The authors also state that the optima identified should fall in the range of sensory intensities represented in the products tested. However, what is done when the optima fall outside of the range of intensities represented is not known, but one could imagine that either the lower or upper boundary is used as the optima.

Westad et al. (2004) published a very interesting manuscript on strategies that can be employed to segment consumers. The researchers describe a general method of fuzzy clustering, an alternative to hierarchical cluster analysis. With fuzzy clustering, the membership to one cluster is not absolute. Instead, each consumer is given a membership value between 0 and 1, relative to each cluster. Westad et al. recommend assigning consumers to a cluster for membership values greater than 0.6. Westad et al. also discuss several strategies to perform clustering. They suggest performing clustering on standardized hedonic data, factor analysis scores of long, skinny preference data, or loadings of the short, fat preference data. They further compare these approaches through the interpretation of the resulting clusters from external data and conclude that clustering of the long, skinny data matrix yielded a better interpretation of the segments based on sensory data. They also proposed the evaluation of cluster stability through a method of random shuffling of preference data.

Vigneau et al. (2001) and Vigneau and Quannari (2002) have done recent work in this field and have proposed interesting methods for consumer segmentation that are based on a clustering-of-variables approach. In this case, consumers are considered variables, and products are observations. The process employed is quite simple in principle. First, an initial clustering solution that consists of K clusters (G_k) of consumers is established, using a hierarchical clustering algorithm that is no different from the techniques described previously, except that a difference criterion is used (i.e., criterion S). Vigneau et al. (2001) complement this initial solution with a partitioning algorithm that allows variables (i.e., consumers) to change their cluster membership to another cluster if the preference scores for a specific consumer are more like those of consumers in another cluster. The latent components (T_k) talked about by Vigneau et al. are defined for each initial cluster as the centroid (the equivalent of the center of gravity for the mathematically impaired) of that cluster:

$$T_k = \frac{\sum_{j=1}^{p_k} x_{kj}}{p_k}, \qquad \text{(Eq. 6.3)}$$

where x_{kj} is the jth centered variable in cluster G_k and p_k is the number of consumers in cluster G_k.

The merging of clusters resulting from hierarchical cluster analysis is done so that the sum of the correlations for all clusters and consumers between the centered consumer data and their respective centroid is maximized. This is referred to criterion S:

$$S = \sum_{k=1}^{K} \sum_{j=1}^{p_k} r(x_{kj}, T_k), \qquad \text{(Eq. 6.4)}$$

where K is the number of clusters, x_{kj} is the jth centered variable in cluster G_k, p_k is the number of consumers in cluster G_k, T_k is the centroid of cluster G_k, and $r(x_{kj}, T_k)$ is the correlation between x_{kj} and T_k.

On the basis of the initial hierarchical cluster solution, the membership of each consumer in a cluster is reassessed by calculating the correlation between the centered preference scores and the centroid scores for the various clusters. The consumer in question is moved to the cluster for which that correlation is maximum. This process is repeated for each consumer and iterated until cluster membership stability is obtained.

Criterion S is also used to determine an appropriate number of clusters for the data. The authors describe a screeplot of S versus the number of clusters. This plot is similar in principle and interpretation to that of residual Y variance versus number of principal components or factors in multivariate regression, which allows us to determine the optimal number of factors to be used in the model. One is looking for a jump in the aggregation criterion when going from a solution with n clusters to a solution with $n + 1$ clusters. This is indicative of the fact that unnatural clusters are being formed.

Sahmer et al. (2004), who is from the same research group, presented an alternative to determine the optimal number of clusters. They perform a hierarchical cluster analysis based on criterion Q:

$$Q = \frac{1}{n} \sum_{k=1}^{K} \sum_{j=1}^{p} \delta_{jk} \|z_j - c_k\|^2, \qquad \text{(Eq. 6.5)}$$

where z_j is standardized scores for consumer j, c_k is the latent variable for cluster k, and δ_{jk} is one if consumer j belongs to cluster k and zero otherwise. The researchers proposed clustering based on the eigenvalues of the matrix $E'E$, where E consists of the residuals between standardized consumer scores and corresponding latent variables. The null hypothesis is that E is formed by noise whereas Ha is that E is structured. The authors define T as

$$T = \frac{\left(\prod_{i=1}^{r} \lambda_i\right)^{1/r}}{\frac{1}{r}\sum_{i=1}^{r} \lambda_i}, \qquad \text{(Eq. 6.6)}$$

where λ_i is the ith eigenvalue of $E'E$ and r is the minimum between $n - 1$ and $p - 1$. The variable T is a value between zero and one, with the closer it is to one, the more likely H_0 is true. The p value associated with T is then estimated through a bootstrap resampling procedure.

References

Helgesen, H., R. Solheim, and T. Naes. 1997. Consumer preference mapping of dry fermented lamb sausages. Food Q. Pref. 8:97–109.

Jaeger, S.R., C.M. Lund, K. Lau, and F.R. Harker. 2003. In search of the "ideal" pear (pyrus spp.): Results of a multidisciplinary exploration. J. Food Sci. 68:1108–1117.

Martinez, C., M.J. Santa Cruz, G. Hough, and M.J. Vega. 2002. Preference mapping of cracker type biscuits. Food Q. Pref. 13:535–544.

McEwan, J.A., P.J. Earthy, and C. Ducher. 1998. Preference mapping: A literature review. Campden & Chorleywood Food Research Association Review No. 6.

McEwan, J.A. 1998. Cluster analysis and preference mapping. Campden & Chorleywood Food Research Association Review No. 12.

Moskowitz, H.R. and R. Bernstein. 2000. Variability in hedonics: Indications of worldwide sensory and cognitive preference segmentation. J. Sensory Stud. 15:263–284.

Moskowitz, H.R., B.E. Jacobs, and N. Lazar. 1985. Product response segmentation and the analysis of individual differences in liking. J. Food Q. 8:168–191.

Popper, R., J. Kroll, and J. Magidson. 2004. Application of latent class models to food product development: A case study. Sawtooth Software Conference Proceedings, 2004.

Sahmer, K., E. Vigneau, and E.M. Qannari. (2004) A cluster approach to analyze preference data: Choice of number of clusters. Proceedings of the 7th Sensometrics Meeting, Davis, CA.

Vermunt, J.K. and J. Magidson (2000). Latent class cluster analysis. In J.A. Hagenaars and A.L. McCutcheon (eds.). Advances in latent class models. Cambridge: Cambridge University Press.

Vigneau, E., E.M. Quannari, P. Punter, and S. Knoops. 2001. Segmentation of a panel of consumers using clustering of variables around latent directions of preference. Food Q. Pref 12:359–363.

Vigneau, E., and E.M. Quannari. 2002. Segmentation of consumers taking into account of external data: A clustering of variables approach. Food Q. Pref. 13:515–552.

Ward, J. H. (1963). Hierarchical grouping to optimize an objective function. J. Am. Stat. Assoc. 58:236–244.

Westad, F., M. Herleth, and P. Lea. 2004. Strategies for consumer segmentation with applications on preference data. Food Q. Pref. 15(2004):681–687.

7 Ordinal Logistic Regression Models in Consumer Research

7.1. Introduction

Consumer data are often of an ordinal nature. Discrete ordinal scales are often used to measure consumer liking, perceived attribute intensity, or attribute level appropriateness (Just About Right) in foods. For example, the commonly used 9-point hedonic scale to assess liking is an ordinal discrete scale, including nine ordered categories (e.g., dislike extremely, dislike very much, dislike moderately, dislike slightly, neither dislike nor like, like slightly, like moderately, like very much, and like extremely). Thus, data obtained using the hedonic scale are ordinal because their order is meaningful. Ordinal data like this are quite often rerepresented by integer numbers (such as 1, 2, ..., 9) and then analyzed as if they were continuous interval level data using analysis of variance. This implies that intervals between two adjacent categories are equally spaced and that the corresponding responses are normally distributed. However, the data obtained using the hedonic or Just About Right scale is essentially categorical and measured at the ordinal, not interval, level. As a consequence, a much better way to deal with such an ordinal response variable is to treat it as an ordinal categorical variable coming from a multinomial distribution. The ordinal nature of the categories is then taken into account by imposing particular constraints on the odds of responding; that is, choosing one category rather than another. As will be further explained below, ordinal data can be analyzed by ordinal logistic regression models such as the proportional odds model (Agresti, 2002).

Logistic regression models are widely used in medical science and marketing (Agresti, 2002; Allison, 2001). The proportional odds model (POM), one of logistic regression models, has recently been applied to sensory data from studies on consumer acceptance of canola oil (Vaisey-Genser et al. 1994), qualitative studies of food choice (Tepper et al. 1997), consumer acceptance of oca cultivars (Sangketkit et al. 2000), and preference mapping (Meullenet et al. 2002). In the following sections of this chapter, odds and odds ratio and binary logistic regression are discussed first, followed by ordinal logistic regression or POM.

7.2. Limitations of Ordinary Least Squares Regression

Linear regression, usually referred to ordinary least squares regression (OLSR), is the simplest form of regression, which is widely used in sensory science for continuous dependent variables. The underlying assumptions on OLSR include the following:

- Independent variables (Xs) can be either continuous or categorical, whereas the dependent variable (Y) must be continuous.
- The dependent variable (Y) or random error term (ε) is normally distributed

- The dependent variable (Y) is measured independently; that is, there are no correlations between separate measurements of the dependent variable (Y).
- The variance of the dependent variable (Y) is constant.

If all assumptions are satisfied, ordinary least square estimates of model parameters (such as regression coefficients and their standard errors) are the best linear unbiased estimator and have minimum sampling variance. If one of the assumptions is violated, ordinary least square estimates (especially standard errors) may be biased, resulting in biased or misleading test statistics. To model the outcomes of a categorical dependent variable such as overall liking scores measured on a hedonic scale, logistic regression is a better alternative. Logistic regression is very similar to linear regression and is now widely used for modeling in social sciences and marketing research. The key difference is that the dependent variable (Y) in logistic regression is not continuous but is discrete or categorical. For discrete or categorical response variables, it may be inappropriate to use OLSR because the response values are not measured on a ratio scale and the error terms may not be normally distributed. In addition, the OLSR model can generate as predicted values any real number ranging from negative to positive infinity, whereas a categorical variable can only take on a limited number of discrete values within a specified range.

7.3. Odds, Odds Ratio, and Logit

To appreciate the logistic model, it is very helpful to have an understanding of odds, log odds (i.e., logit), and odds ratio. It is generally believed that probability is a natural way to quantify the chances that an event will occur. We automatically think in terms of probabilities ranging from 0 to 1, with 0 meaning that the event will certainly not occur and 1 meaning that the event certainly will occur (Allison, 2001). For professional gamblers, however, the odds are another natural way of presenting probabilities or chances of events. The odds of an event happening is the ratio of probability that the event will happen to the probability that the event will not happen; namely,

$$\text{odds} = \frac{\text{probability of event}}{\text{probability of no event}} = \frac{p}{q} = \frac{p}{1-p};$$

$$p = \frac{\text{odds}}{1+\text{odds}}, \qquad \text{(Eq. 7.1)}$$

where p is the probability of an event happening (generically called success) and $q = 1 - p$ is the probability of an event not happening (generically called failure). Calculation of odds is straightforward and simple. For example, if success probability $p = 0.8$ and failure probability $q = 1 - p = 0.2$, then the odds of success are calculated as follows:

$$\text{odds (success)} = p/q = 0.8/0.2 = 4.0; \qquad \text{(Eq. 7.2)}$$

that is, the odds of success are 4 to 1, meaning that out of five trials, there will be four successes and one failure. Similarly, the odds of failure would be

$$\text{odds (failure)} = q/p = 0.2/0.8 = 0.25. \qquad \text{(Eq. 7.3)}$$

This means that the odds of failure are 1 to 4. The odds of success and the odds of failure are just reciprocals of one another (i.e., 1/4 = .25 and 1/.25 = 4). If we know the odds, the corresponding probabilities can be also easily calculated. The relationship between probability and odds is illustrated in Table 7.1. The table shows that odds less than 1.0 correspond to probabilities below 0.5, whereas odds greater than 1.0 correspond to probabilities over 0.5. Similar to probability, the low bound of odds is 0. The upper bound of odds is $+\infty$, which is quite different from the upper bound (1.0) of probability.

It is often more convenient to work with the log of the odds than with the odds themselves. The log base e (log) of the odds is therefore defined as the logit or log odds; namely,

$$\text{logit}(p) = \log(\text{odds}) = \log\left(\frac{p}{q}\right) = \log\left(\frac{p}{1-p}\right). \quad \text{(Eq. 7.4)}$$

The relationship between the odds and logit is also presented in Table 7.1. The table shows that the range of the logit covers $-\infty$ (low bound) to $+\infty$ (upper bound), which is a good property of the logit on which the ordinary least square regression is used to derive logistic regression.

Odds is a sensible scale for multiplicative comparisons (Allison, 2001) and is widely used to calculate an odds ratio, which is defined as the ratio of the odds of an event occurring in one group to the odds of it occurring in another group. These groups might be men and women, an experimental group and a control group, or any other dichotomous classification. If odds_1 and odds_2 are the odds for groups 1 and 2, respectively, the odds ratio is defined as

$$\text{odds ratio} = \frac{\text{odds}_1}{\text{odds}_2}. \quad \text{(Eq. 7.5)}$$

The odds ratio is a useful measure of association for a variety of study designs. When two odds are equal, the odds ratio is one, indicating that two groups have similar association with the response variable. When two odds are not equal, the odds ratio is either greater or less than one, indicating that one group has stronger or weaker association with the response variable than another group. Because of its ease of interpretation, the odds ratio is often used to interpret the results from logistic regression. An example from a consumer preference test with 150 consumers evaluating two products, A and B,

Table 7.1. Relationship between probability, odds, and logit.

Probability		Odds		Logit	
Success	Failure	Success	Failure	Success log	Failure log
p	$q = 1 - p$	$p/(1-p)$	$(1-p)/p$	$(p/[1-p])$	$([1-p]/p)$
0.0	1.0	0.00	$+\infty$	$-\infty$	$+\infty$
0.1	0.9	0.11	9.00	−0.954	0.954
0.2	0.8	0.25	4.00	−0.602	0.602
0.3	0.7	0.43	2.33	−0.368	0.368
0.4	0.6	0.67	1.50	−0.176	0.176
0.5	0.5	1.00	1.00	0.000	0.000
0.6	0.4	1.50	0.67	0.176	−0.176
0.7	0.3	2.33	0.43	0.368	−0.368
0.8	0.2	4.00	0.25	0.602	−0.602
0.9	0.1	9.00	0.11	0.954	−0.954
1.0	0.0	$+\infty$	0.00	$+\infty$	$-\infty$

Table 7.2. Preference by gender in a paired preference test.

Response	Men	Women	Total
Preferred A	31	19	50
Preferred B	39	61	100
Total	70	80	150

is given in Table 7.2. Overall, the estimated odds of "preferred A" responses are 50/150 = 0.333 or 1 to 3. In other words, on average, one of four customers preferred product A, whereas three consumers did not. For men, the odds are 31/39 = 0.795. For women, the odds are 19/61 = 0.311. The ratio of the men odds to the women odds is 0.794/0.311 = 2.552. The odds of men preferring product A is 155.2% greater than for women; the odds of women preferring A are 1/2.552 = 0.392 times the odds of men preferring A. Just by inversing the odds of preferring A, the odds of preferring B—or not preferring A—for men, women, and all consumers are obtained as 1/0.794 = 1.258, 1/0.311 = 3.211, and 1/.333 = 3.00, respectively. The odds ratio of the women odds of preferring B to the men odds is 2.552, indicating that the odds of women preferring B is 155.2% greater than for men. The results show that more men and women preferred B and that in terms of the odds, 155.2% more women preferred B than men, whereas 155.2% more men preferred A. Significance tests for the odds ratio can be done using modern statistical programs. SAS codes for this example follow. SAS output partially given here shows that the odds ratio is 2.552, which is significantly different from 1 because the 95% confidence limits (1.2015, 5.4668) do not contain 1. At the significance level of 5%, gender has a significant effect on preference, which can be obtained using the binary logistic regression discussed in the following section.

```
DATA Table72;
     INPUT Prod$ Sex$ Count @@;
     Pref=(Prod='A');           *for binary logistic model;
     DATALINES;
     A    Men       31    A    Women      19
     B    Men       39    B    Women      61
     ;
RUN;

PROC FREQ DATA=Table72;
     WEIGHT Count;
     TABLES Prod*Sex;
     EXACT odds ratio;          *exact test for Odds Ratio;
RUN;

Partial SAS Output
          Odds Ratio (Case-Control Study)

          Odds Ratio                          2.5520

          Asymptotic Conf Limits
          95% Lower Conf Limit                1.2695
          95% Upper Conf Limit                5.1301
```

```
Exact Conf Limits
95% Lower Conf Limit         1.2015
95% Upper Conf Limit         5.4668

         Sample Size = 150
```

7.4. Binary Logistic Regression

To understand ordinal logistic regression, one needs to understand the binary regression. For simplicity, consider the situation in which a food product is evaluated by 80 consumers in an effective test without replications. For each consumer, there are only two possible outcomes of the response: liked and disliked. By convention, a value of 1 is used to indicate success ("liked" in this case), and a value of 0 is used to signify failure ("disliked" here). This situation also arises frequently in discrimination tests, in which consumers make either correct or incorrect choices, and in preference tests, in which consumers prefer or do not prefer a product. In general, a binary random variable can be defined as

$$Z = \begin{cases} 1 & \text{if outcome is a success} \\ 0 & \text{if outcome is a failure} \end{cases} \quad \text{(Eq. 7.6)}$$

with probabilities $P(Z = 1) = \pi$ (the underlying probability of success) and $P(Z = 0) = 1 - \pi$ (the probability of failure). For n consumers, there are n random response variables $Z_1, Z_2, \ldots, Z_n$. Each random response variable (i.e., the response of each consumer) is assumed to follow the Bernoulli distribution in each trial. In addition, there are a number of explanatory variables or covariates ($x_{i1}, x_{i2}, \ldots, x_{ip}$) associated with each response. In consumer testing, for example, the covariates may be consumers' gender, age, family income, number of family members, status of employment, race, and so on, depending on the research objectives. Testing sessions, testing days, sample preparations, and other factors can also be treated as covariates. For consumer testing, results from individual consumers are often represented in a table format such as Table 7.3, which lists consumer ID, response, and other information such as gender and age group. The effects of covariates such as gender and age group on the response can be analyzed by binary logistic regression. Before binary logistic regression is applied to the data in Table 7.3, however, let us explain the difference between grouped and ungrouped data.

It is important to distinguish between grouped and ungrouped binary or categorical data because logistic regression is based on grouped data. Table 7.3 is a typical example of ungrouped data that are listed by individual consumers. Ungrouped data can be grouped by covariates of interest. Using these covariates (gender and age group), we can easily convert the ungrouped data (Table 7.3) into grouped data (Table 7.4). The conversion of the ungrouped to the grouped is automatically done by computer software programs before fitting logistic regression models. Such a table (Table 7.4) is often called the contingency table. The contingency table (Table 7.4) shows that there are 12 distinct samples or populations. Each sample is one of the 12 combinations of the covariates gender and age groups (i.e., 2 for gender × 6 for age group). Number of successes Y_i for the ith sample is the sum of response Z_j ($j = 1, 2, \ldots, n_i$) in the sample; namely,

Table 7.3. Individual liking/disliking results in a consumer test.

Consumer ID	Response	Gender	Age Group
1	0	0	3
2	1	1	2
3	1	0	1
4	1	0	2
5	1	0	4
6	1	0	3
7	0	1	5
8	1	0	5
9	1	1	5
10	1	0	5
...	...	...	...
79	1	0	4
80	0	0	5

Note: 1 = liked and 0 = disliked for response; 1 = male and 0 = female for Gender; 1 = 18–20, 2 = 21–23, 3 = 24–26, 4 = 27–29, 5 = 30–32, and 6 = 33–35 for Age Group.

Table 7.4. Contingency table of consumer data.

Sample	Gender	Age Group	No. of Disliked	No. of Liked (Y)	Total (n)	Observed P_i
1	0	1	0	7	7	1.000
2	0	2	1	10	11	0.909
3	0	3	2	5	7	0.714
4	0	4	1	7	8	0.875
5	0	5	1	4	5	0.800
6	0	6	0	9	9	1.000
7	1	1	0	2	2	1.000
8	1	2	1	6	7	0.857
9	1	3	0	2	2	1.000
10	1	4	1	6	7	0.857
11	1	5	2	5	7	0.714
12	1	6	0	8	8	1.000
Total			9	71	80	0.888

Note: 1 = liked and 0 = disliked for response; 1 = male and 0 = female for Gender; 1 = 18–20, 2 = 21–23, 3 = 24–26, 4 = 27–29, 5 = 30–32, and 6 = 33–35 for Age Group.

$$Y_i = \sum_{j=1}^{n_i} Z_j \quad (i = 1, 2, \ldots, I), \qquad \text{(Eq. 7.7)}$$

which has a binomial distribution $Y_i \sim \text{binomial}(n_i, \pi_i)$, where n_i is the number of consumers in the ith sample, π_i is the probability of success in the ith sample, and I is the number of samples, $0 \leq Y_i \leq n_i$ and $n_1 + n_2 + \ldots + n_I = n$. The expected mean $E(Y_i)$ and variance $Var(Y_i)$ are given by

$$\begin{aligned} E(Y_i) &= n_i \pi_i; \\ Var(Y_i) &= n_i \pi_i (1 - \pi_i) \quad (i = 1, 2, \ldots, I). \end{aligned} \qquad \text{(Eq. 7.8)}$$

Similar to other categorical models, logistic regression models use contingency tables or grouped data by covariates. A nice property of grouped data is that the response can be represented in the probability form of event/trial (i.e., the proportion of success $P_i = Y_i/n_i$). Ungrouped data can be considered as a special case of grouped data, with $n_1 = n_2 = \ldots = n_I = 1$. The expected mean of the proportion of success is $E(P_i) = \pi_i$, which simply is the probability of success. For $i = 1, 2, \ldots, I$ samples, the probability π_i may be modeled using OLSR to examine the effects of covariates on the response; that is,

$$\pi_i = \alpha_0 + \alpha_1 x_{i1} + \alpha_2 x_{i2} + \ldots + \alpha_p x_{ip} + \varepsilon_i \qquad \text{(Eq. 7.9)}$$

where $x_{i1}, x_{i2}, \ldots, x_{ip}$ are observed covariates; $\alpha_0, \alpha_1, \alpha_2, \ldots, \alpha_p$ are the regression parameters; p is the number of covariates; I is the number of populations or samples; and ε_i is a random error term, normally distributed with zero mean and constant variance. The serious problems with this OLSR model are that π_i is a probability and is therefore restricted to the interval [0,1], whereas the fitted values of π_i may sometimes lie outside the interval [0,1], resulting in meaningless probabilities of less than 0 or greater than 1; and that the variance of the proportion of success P_i, an estimate of π_i, is derived as:

$$Var(P_i) = \frac{\pi_i(1-\pi_i)}{n_i}. \qquad \text{(Eq. 7.10)}$$

which depends on the probability π_i itself; that is, the variance increases as π_i increases, whereas OLSR assumes a constant variance for all samples. Because of this, OLSR models may be inappropriate when dealing with a binary response variable. An alternative to OLSR is logistic regression, which ensures that the fitted values of π_i lie within the interval [0,1] through a logistic function transformation. Despite its clearcut violation of assumptions, OLS linear regression actually does a surprisingly good job in most applications of OLS regression to dichotomous variables in that it give results qualitatively quite similar to results obtained using logit regression (Allison, 2001).

This example also shows a common problem encountered in consumer data; that is, some categories have only few observations (e.g., samples 7 and 9 have only two observations), which may be not a large enough sample for valid statistical analysis. This type of problem is often associated with either too many categories or too small a number of consumers being used in tests. A common way to deal with the problem is to collapse adjacent categories to increase the number of observations in each category. For example, Table 7.5 is obtained by combining the original six categories for age group into three categories. By comparing Tables 7.4 and 7.5, we can see that the proportions of liking in

Table 7.5. Combined contingency table of consumer data.

Sample	Gender	Age Group	No. of Disliked	No. of Liked (Y)	Total (n)	Observed P_i
1	0	12	1	17	18	0.944
2	0	34	3	12	15	0.800
3	0	56	1	13	14	0.929
4	1	12	1	8	9	0.889
5	1	34	1	8	9	0.889
6	1	56	2	13	15	0.867
Total			9	71	80	0.888

Note: 1 = liked and 0 = disliked for response; 1 = male and 0 = female for Gender; 12 = 18–23, 34 = 24–29, and 56 = 30–35 for Age Group.

the combined table are more homogeneous, indicating that there are no effects of gender and age group on liking.

As discussed previously, the problem with directly fitting a multiple linear regression model to probabilities is that probabilities are bounded between 0 and 1, whereas the regression model is inherently unbounded ($-\infty, +\infty$). The solution is to transform the probability into log odds (logit) so that it is no longer bounded (Table 7.1). Therefore, by fitting a multiple regression model to the log odds (logit) instead of probability itself, we obtain the following general linear logistic regression model for binary data:

$$\text{logit}(\pi_i) = \log\left(\frac{\pi_i}{1-\pi_i}\right) = \beta_0 + \beta_1 x_{i1} + \beta_2 x_{i2} + \ldots + \beta_p x_{ip} \quad \text{(Eq. 7.11)}$$

This model is often referred to as the logit model. The logit is a transformation that takes a number π_i between 0 and 1 and transforms it to the following:

$$\log\left(\frac{\pi_i}{1-\pi_i}\right), \quad \text{(Eq. 7.12)}$$

which removes the low and upper bounds (0 and 1) of probabilities. Once the model has been fitted, response probability can be calculated as follows:

$$\pi_i = \frac{e^{\beta_0 + \beta_1 x_{i1} + \beta_2 x_{i2} + \ldots + \beta_p x_{ip}}}{1 + e^{\beta_0 + \beta_1 x_{i1} + \beta_2 x_{i2} + \ldots + \beta_p x_{ip}}} \quad \text{(Eq. 7.13)}$$

which is referred to as the logistic function and can also be written in a simpler form as follows:

$$\pi_i = \frac{1}{1 + e^{-(\beta_0 + \beta_1 x_{i1} + \beta_2 x_{i2} + \ldots + \beta_p x_{ip})}} \quad \text{(Eq. 7.14)}$$

The procedure of logistic regression can be summarized as follows. First, data are grouped in terms of covariates of interest to form samples (also called sample profiles) or populations (or population profiles). Second, the probability of "success" (π) is estimated for each sample profile. Third, the log of the odds $\log(\pi/[1-\pi])$ for each sample profile is calculated. Finally, a linear regression model in which $\text{logit}(\pi) = \log[\pi/(1-\pi)]$ is the dependent variable (Y) and the covariates are the independent variables (X) is fitted.

Compared to the linear regression, logistic regression requires extra steps to preprocess the data and transform binary response variables into the logit, mainly because the response variable is categorical. Because logistic regression is very similar to OLSR, it is a robust tool for developing predictive models. Keep in mind that the relationship between the logit and the predictors (X) is linear, but the relationship between the probabilities and the predictors is not.

7.4.1. Estimation of the Logit Model

Regression coefficients of logit models can be estimated using one of three methods—OLS, weighted least squares, and maximum likelihood (ML)—if sample data are grouped in some way. However, OLS and weighted least squares cannot be applied to ungrouped data because the logit transformation cannot be computed for values of 1 and 0 (i.e., $-\infty$ or $+\infty$). ML is the only method that can be used to estimate the logit models for both ungrouped and grouped data (Allison, 2001).

ML is a very general approach to estimation that is widely used for all sorts of statistical models. The basic principle of ML is to choose as estimates those parameter values that, if true, would maximize the probability of observing what we have observed. There are two steps to this (Allison, 2001). First, the probability of the data is expressed as a likelihood function of unknown parameters, and second, the values of the unknown parameters that maximize the likelihood function are determined. Details on ML estimations can be found in Agresti (2003), Lawal (2004), and Allison (2001).

7.4.2. Goodness-of-Fit Test Statistics

Once the logit model has been estimated, the model fit to the data needs to be assessed using so-called goodness-of-fit test statistics. There are a number of goodness-of-fit test statistics available. The goodness of fit of a logistic regression model is often assessed by the Pearson chi-square (χ^2), the log-likelihood ratio (G^2), and the deviance (D). The χ^2 test statistic, introduced by Karl Pearson (1900), is one of the foremost goodness-of-fit test statistics and is given by

$$X^2 = \sum_{i=1}^{I} \frac{(O_i - E_i)^2}{E_i}, \qquad \text{(Eq. 7.15)}$$

where $E_i = \pi_i n_i$ is the expected frequency under the null hypothesis H_o (i.e., the model is correct) for the ith sample, and O_i is the corresponding observed frequency. The Pearson χ^2 statistic has asymptotic χ^2 distribution with $(I-1)$ degrees of freedom under H_o. When there is perfect fit between the null model and the data (i.e., $E_i = O_i$), $\chi^2 = 0$. As the discrepancy increases, χ^2 increases too. When χ^2 exceeds some appropriate upper percentage point of a χ^2 distribution with $(I-1)$ degrees of freedom, H_o is rejected.

The likelihood ratio test statistic G^2, well known as the log-likelihood ratio test, is defined (Wilks, 1938) as:

$$G^2 = 2\sum_{i=1}^{I} O_i \log\left(\frac{O_i}{E_i}\right), \qquad \text{(Eq. 7.16)}$$

where E_i and O_i have the same meanings as in the Pearson χ^2. G^2 is more widely used for categorical data than χ^2 because it has two advantages: the overall G^2 can be decomposed into small components, and it simplifies the process of comparing one model against another (Lawal, 2004).

The deviance (D), first introduced by Nelder and Wedderburn (1972), is written as:

$$D = 2[L_{\max} - L_0], \qquad \text{(Eq. 7.17)}$$

where $L_{\max}$ and L_0 are the log-likelihood functions for the maximal model (also referred to as the saturated model) and the model of interest, respectively. The deviance implicitly involves a comparison between the model of interest and a "maximal" model that is more complex. Because a saturated model has one parameter for each of the combinations of all predictors, it produces a perfect fit to the data. The maximal model always fits better than the model of interest. The question is whether the difference in fit could be explained by chance. The derivation shows that the deviance D is equivalent to the log-likelihood ratio G^2; namely, $D = G^2$ (Lawal, 2004). Under the null hypothesis (H_o) that the model is correct, the following approximation can be used for hypothesis testing.

$$\chi^2 \sim \chi^2(n - p);$$
$$G^2 \sim \chi^2(n - p);$$
$$D^2 \sim \chi^2(n - p). \hspace{2cm} \text{(Eq. 7.18)}$$

The χ^2 is asymptotically equivalent to G^2 or D. The choice between G^2 or D and χ^2 depends on the adequacy of the approximation to the $\chi^2(n - p)$ distribution. The variables χ^2, G^2, and D do not have a χ^2 distribution with ungrouped data. As a result, there are no p-values available for χ^2, G^2, and D when working with ungrouped data.

7.4.3. Model Diagnostics

The model adequacy can be evaluated using two forms of residuals: the deviance residual and the Pearson residual. There are also standardized versions of each of these types of residuals. Most statistical packages provide various residuals for checking model adequacy. Obviously, the residuals are useful in determining which observations are most poorly fitted by the model.

7.4.4. Overdispersion

Overdispersion is another important issue that must be given consideration when assessing the adequacy of models for binary response. Overdispersion refers to the situation where the observed binary data (Y_i) have a variance larger than that expected under the binomial model (i.e., $\text{var}[Y_i] > n_i\pi_i[1 - \pi_i]$). This phenomenon is quite common in practice. There are two possible reasons: inadequate model and correlated data. One approach to deal with overdispersion is to include an extra parameter θ in the model to account for overdispersion (i.e., $\text{var}[Y_i] > n_i\pi_i[1 - \pi_i]\varphi$).

7.4.5. Multicollinearity

Multicollinearity occurs in any regression models (e.g., logit) when there are strong linear dependencies among all or some of the explanatory variables. Although no universally acceptable definition of multicollinearity has been established, correlations of 0.70 and above are frequently mentioned as benchmarks (Rud, 2001). The major issue about multicollinearity is that it increases the standard error of the sampling distribution of the coefficients of highly correlated variables. However, multicollinearity does not violate any of the assumptions of OLS regression, and thus the OLS parameter estimator under such circumstances is still the best linear unbiased estimator (Achen, 1982; Rud, 2001). When explanatory variables in the model are highly correlated with each other, statistical power can substantially drop because the amount of variation held in common among the explanatory variables can leave little remaining data to reliably estimate the separate effects of each. When some variables have little or weak effects on the response variable individually but have strong effects as a group, it indicates multicollinearity. If the model is used for interpretive purpose, then multicollinearity is a serious problem because the effects of each explanatory variable on the response variable may not be interpretable.

Although multicollinearity causes difficulty in allocating unbiased explanatory power to each variable, it is unlikely that it affects model predictions because of the unbiased parameters. In addition, not all variation between two predictors is redundant. By deleting a highly correlated variable, we actually run the risk of throwing away additional useful predictive information, such as the independent or unique variation accounted for by the deleted predictor, resulting in a poorer predictive model. This indicates that multicollinearity is not harmful in the context of models generated for predictive purposes (Rud, 2001).

The problem is then to determine an appropriate course of action when multicollinearity is present. A common solution is to delete one or more offending collinear variables or to use factor or principal components analysis to reduce the amount of redundant variations presented in data. In sensory evaluation, as there are strong correlations among all or some explanatory variables, principal components analysis or partial least squares regression is usually used to avoid multicollinearity problems.

7.4.6. Examples for Binary Logistic Model

The data in Table 7.2 are used as a simple example of applications of binary logistic models. Let π be the probability of Pref = 1 (i.e., preferring A). The odds and logit are $\pi/(1 - \pi)$ and $\log(\pi/[1 - \pi])$, respectively. If the purpose is to test whether the probability of preferring A (Pref = 1) is significantly different from 0.50, the following logistic regression model can be used without any predictors:

$$\text{logit}(\pi) = \log\left(\frac{\pi}{1-\pi}\right) = \beta_0 \quad \text{(Eq. 7.19)}$$

If β_0 is significantly different from zero at a significance level α, it indicates that the probability of preferring A is significantly greater or less than 0.50, depending on the sign of β_0. If the sign of β_0 is negative, the probability of preferring A is significantly less than 0.5; otherwise, the probability of preferring A is significantly greater than 0.5. This model can be fitted by the SAS package. SAS provides four procedures to fit logit models using the maximum likelihood method: LOGISTIC, PROBIT, GENMOD, and CATMOD. In this example, either LOGISTIC or GENMOD is sufficient. The SAS code using LOGISTIC to fit a logit model is given here:

```
PROC LOGISTIC DATA=Table72 DESCENDING ALPHA=0.05;
    WEIGHT Count;
    MODEL Pref=;
RUN;
```

By default, the LOGISTIC procedure arranges the values of the dependent variable from lowest to highest and predicts the lowest value. To predict the highest value of the response variable, the DESCENDING option needs to be used. In this case, the DESCENDING option is used to predict the probability of preferring A (i.e., Pref = 1). If the DESCENDING option had been omitted, the result would be a logit model predicting the probability of preferring B (i.e., Pref = 0). Similarly, a different significance level α can be specified through the ALPHA option. There are numerous options available in the LOGISTIC procedure, and more information can be obtained from SAS online documents.

After submitting the above SAS code, SAS estimates a logit model using the maximum likelihood method. The SAS output is given below.

```
The LOGISTIC Procedure

                    Model Information
    Data Set                         WORK.TABLE7.2
    Response Variable                Pref
    Number of Response Levels        2
    Weight Variable                  Count
    Model                            binary logit
    Optimization Technique           Fisher's scoring

    Number of Observations Read      4
    Number of Observations Used      4
    Sum of Weights Read              150
    Sum of Weights Used              150

                    Response Profile
         Ordered           Total          Total
          Value    Pref  Frequency       Weight
            1       1        2          50.00000
            2       0        2         100.00000

            Probability modeled is Pref = 1.

Model Convergence Status Convergence criterion (GCONV = 1E-8)
                         satisfied.
                   -2 Log L = 190.954

    Analysis of Maximum Likelihood Estimates Standard Wald
    Parameter  DF  Estimate   Error   Chi-Square  Pr > ChiSq
    Intercept   1  -0.6931   0.1732    16.0151     <.0001
```

It can be seen from the SAS output that the model information is given first, followed by response profile, model convergence status, and analysis of maximum likelihood estimates. If the convergence criterion for maximum likelihood method is not satisfied, the corresponding results may be invalid. In the example, the deviance (−2 Log L) is 190.954, which is very useful when comparing two models but does not have much meaning in this example. The "Analysis of Maximum Likelihood Estimates" is the heart of the output. It gives coefficient estimates, their estimated standard errors, and test statistics for the null hypotheses that each coefficient is equal to zero. The Wald χ^2, sometimes called squared z or t statistic, is the test statistic, which is calculated by dividing each coefficient by its standard error and squaring the result. The estimate for the intercept is −0.6931, which is highly significantly different from zero at $\alpha = 0.05$ by Wald χ^2. This indicates that the probability of preferring A ($\pi = 1/[1 + \exp(0.6931)] = 0.3333$; see Equation 7.14) is significantly less than 0.5. There is no advantage of logistic regression over more traditional methods (e.g., a binomial formula or table) to test preference significance for this simple example. However, if one needs to examine the effects of various factors (such as sex, age and other factors) on preference, logistic regression has advantages. For example, if the purpose is to evaluate the effect of gender on the probability of preferring A (Pref = 1; Table 7.2), the following logistic regression model with one predictor (gender) can be used:

$$\text{logit}(\pi) = \log\left(\frac{\pi}{1-\pi}\right) = \beta_0 + \beta_1 * Gender. \qquad \text{(Eq. 7.20)}$$

SAS codes for this model, in which some new options have been used, are given below. The CLASS statement is used to declare categorical variables, whereas the variables that are not listed in the CLASS statement are continuous. SAS will create dummy variables for each categorical variable. There are various coding schemes for creating dummy variables. In the code we have used, the PARAM = GLM option to require the GLM dummy variable coding scheme for Gender, as used in PROC GLM. If the PARAM option is omitted, the default coding scheme is used. AGGREGATE is used to ask for the computation of the goodness-of-fit statistics, whereas the SCALE = none option means to not adjust the goodness-of-fit statistics for overdispersion. Although AGGREGATE is useful when there are a few categorical variables involved, when there are too many variables or some continuous variables, AGGREGATE is no longer useful because there will be nearly as many profiles as original observations, and nothing will be accomplished by aggregating (Allison, 2001). RSQUARE is requested for the likelihood-based pseudo-R^2 measurement for the model, which mimics the coefficient of determination R^2. However, it cannot be interpreted as a proportion of variance "explained" by the independent variables (Allison, 2001). Because the upper bound of the pseudo-R^2 is less than 1, it is better to use rescaled pseudo R^2 (labeled the "Max-rescaled Rsquare" in PROC LOGISTIC).

The SAS output usually gives three model fit statistics: AIC (Akaike's information criterion), SC (Schwartz criterion), and −2 Log L. AIC and SC are useful in comparing the relative fit of different models, and in general, lower values of these statistics correspond to better models. The deviance, −2 Log L, is the model fit statistic used throughout this chapter. In this example, all three statistics indicate that the logit model is better than the model with intercept only. Max-rescaled pseudo-R^2 is 0.8314, indicating that the logit model fits adequately well. In the section of "Testing Global Null Hypothesis: BETA = 0," there are three χ^2 statistics: Likelihood Ratio, Score, and Wald. Likelihood ratio can be calculated from the difference in deviance between two models; that is, 7.1210 = 190.954 − 183.833 in this case. In large samples, all three tests will produce quite similar results. In small samples or samples with extreme data patterns, the likelihood ratio χ^2 is superior to the other two statistics (Jennings, 1986). According to the likelihood ratio statistic and its p-value of 0.0076, the global null hypothesis (BETA = 0) is rejected at $\alpha = 0.05$, and we conclude that at least one of the coefficients is not equal to zero.

The maximum likelihood estimate of the intercept is −1.1664, which is the estimated logit when Gender = Women. The corresponding odds are exp(logit) = exp(−1.1664) = 0.3115. The coefficient for variable Gender = Men is 0.9369, which means that for a 1-unit increase in Gender (changing from women to men), the expected change in log of odds is 0.9369. The odds for men are exp(−1.1664 + 0.9369 × 1) = 0.7949. The ratio of the odds of a male preferring A to the odds of a female preferring A are 0.7949/0.3115 = 2.5521, which is significant at $\alpha = 0.05$, as 95% Wald confidence limits in the SAS output indicate. In conclusion, men were more likely to prefer product A than women.

```
PROC LOGISTIC DATA=Table72 DESCENDING ALPHA=0.05;
    WEIGHT Count;
    CLASS gender/PARAM=glm;
    MODEL Pref=gender/AGGREGATE RSQUARE;
RUN;
```

```
                      Model Fit Statistics
    Criterion    Intercept Only    Intercept and Covariates
        AIC          192.954           187.833
        SC           192.341           186.606
        -2 Log L     190.954           183.833
    R-Square   0.8314      Max-rescaled R-Square   0.8314

        Testing Global Null Hypothesis: BETA = 0
    Test                    Chi-Square   DF    Pr > ChiSq
    Likelihood Ratio          7.1210      1      0.0076
    Score                     7.0848      1      0.0078
    Wald                      6.9152      1      0.0085

               Type 3 Analysis of Effects
    Effect   DF   Wald   Chi-Square   Pr > ChiSq
    Sex       1          6.9152         0.0085

    Analysis of Maximum Likelihood Estimates   Standard   Wald
    Parameter    DF   Estimate   Error    Chi-Square   Pr > ChiSq
    Intercept     1   -1.1664    0.2627    19.7113      <.0001
    Sex Men       1    0.9369    0.3563     6.9152       0.0085
    Sex Women     0    0          .           .            .

                   Odds Ratio Estimates
              Effect       Point Estimate      95% Wald
                                             Confidence Limits
Sex Men vs Women    2.552      1.269          5.130
```

The logit model can be used with multiple explanatory variables. By considering the data in Tables 7.4 and 7.5 (downloadable from our Web site), the probability (π) of liked = 1 can be modeled using multiple predictors by the following binary logistic regression model with SAS codes. The LACKFIT option is used for performing the Hosmer-Lemeshow test, which was proposed by Hosmer and Lemeshow (2000) to remedy the deficiency of AGGREGATE when many variables or continuous variables are involved (please note that LACKFIT is used here for demonstration purposes because only two categorical variables are used in the model and AGGREGATE is still valid):

$$\text{logit}(\pi) = \log\left(\frac{\pi}{1-\pi}\right) = \beta_0 + \beta_1 * Gender + \beta_2 * AgeGroup. \qquad (\text{Eq. 7.21})$$

```
*Import data set in Table 7.4;
DATA TABLE74;
    INFILE   "G:\Book  Data\Chapter7-Table74.csv"    DLM=","
FIRSTOBS=2;
```

```
        INPUT ProdID$ PanID Overall Gender$ AgeGroup Liked;
RUN;

*Combine some age groups to obtain Table 7.5;
DATA TABLE75;
    SET TABLE74;
    IF AgeGroup=1 or AgeGroup=2 THEN CombAgeGroup=12;
    IF AgeGroup=3 or AgeGroup=4 THEN CombAgeGroup=34;
    IF AgeGroup=5 or AgeGroup=6 THEN CombAgeGroup=56;
RUN;

*Fit logit mode using the original Age Group;
PROC LOGISTIC DATA=TABLE74 DESCENDING ALPHA=0.05;
    CLASS Gender AgeGroup/PARAM=glm;
    MODEL Liked=Gender AgeGroup/AGGREGATE RSQUARE LACKFIT;
RUN;

*Fit a logit model using the combined AgeGroup;
PROC LOGISTIC DATA=TABLE75 DESCENDING ALPHA=0.05;
    CLASS Gender CombAgeGroup/PARAM=glm;
    MODEL   Liked=Gender   CombAgeGroup/AGGREGATE   RSQUARE
LACKFIT;
RUN;
```

As we discussed earlier, SAS estimates a logit model using the maximum likelihood method. The maximum likelihood method usually converges after several iterations, but this is not always the case. When all categories in Table 7.4 were used, SAS gave a warning that the maximum likelihood estimate may not exist and that validity of the model fit was questionable (output not shown). Too few observations in some categories may cause the estimation failure. When some categories were combined (Table 7.5), however, the maximum likelihood estimate existed. The SAS output shows that the global null hypothesis (BETA = 0) is not rejected, and neither gender nor age group significantly affected the liking (liked = 1), which is in agreement with the earlier results by odds ratio. The Hosmer and Lemeshow test shows that the null model fits the data. As Allison (2001) pointed out, the Hosmer and Lemeshow test is not a very powerful test, although it is popular.

```
            Binary Logistic Regression For Table 7.5

        Testing Global Null Hypothesis: BETA = 0
        Test              Chi-Square   DF   Pr > ChiSq
        Likelihood Ratio  1.1398       3    0.7675
        Score             1.1705       3    0.7601
        Wald              1.1282       3    0.7703

                  Type 3 Analysis of Effects
        Effect          DF   Wald Chi-Square   Pr > ChiSq
        Gender          1    0.0425            0.8367
        CombAgeGroup    2    1.0934            0.5789
```

```
          Partition for the Hosmer and Lemeshow Test
               Liked = 1                   Liked = 0
    Group    Total    Observed    Expected    Observed    Expected
      1        9         8          7.38         1         1.62
      2       15        12         12.62         3         2.38
      3       15        13         13.35         2         1.65
      4       14        13         12.65         1         1.35
      5        9         8          8.27         1         0.73
      6       18        17         16.73         1         1.27

          Hosmer and Lemeshow Goodness-of-Fit Test
               Chi-Square    DF    Pr > ChiSq
                 0.8318       4      0.9341
```

7.5. Ordinal Logistic Regression Models

As discussed above, binary logistic regression is a very useful tool for analyzing responses having two categories. These binary responses often come from consumer discrimination or preference tests. However, for consumer acceptance tests, the responses are often ordinal. Ordinal responses are common in consumer testing, marketing research, and opinion polls. When the response categories have a natural ordering, model specification should take that into account so that extra information is used in the model. There are two reasons for preferring models that take the ordering into account (Allison, 2001): they are much easier to interpret, and hypothesis tests are more powerful. However, ordered models impose restrictions on the data that may be inappropriate. Therefore, whenever using an ordered model, it is important to test whether its restrictions are valid (Allison, 2001). There are three ordinal logistic regression models that take account of such ordinality: the proportional odds model, the adjacent categories model, and the continuation ratio model.

All three models reduce to the binary logit models when the response variable has only two categories. These models can be considered as extension of the binary logit model; that is, they fit multiple simultaneous binary logits to ordinal responses. The continuation ratio model is designed for situations in which the ordered categories represent a progression through stages. There are few situations for applications of the continuation ratio model in sensory science, so this model will not be discussed. Despite significant differences between the proportional odds model and the adjacent categories model, the two models tend to yield very similar conclusions in practice (Allison, 2001). Therefore, only the proportional odds model will be discussed in the following section.

7.6. Porportional Odds Model (POM)

Suppose that an ordinal response variable Y has J categories with a natural ordering: $Y_1 < Y_2 < Y_3 < \ldots < Y_J$ or $Y_1 > Y_2 > Y_3 > \ldots > Y_J$. A good example of ordinal categories used in consumer testing is the popular 9-point hedonic scale: dislike extremely, dislike very much, dislike moderately, dislike slightly, neither dislike nor like, like slightly, like moderately, like very much, and like extremely. The hedonic scale has a natural ordering from disliking extremely to liking extremely. The ordinal categorical response from a 9-point hedonic scale is traditionally analyzed by assuming equal intervals between categories and assigning numeric numbers from 1 to 9 to categories in natural order (i.e., dislike extremely = 1, dislike very much = 2, . . . , like very much = 8, and like extremely = 9). The assumption about equal intervals between categories may be not always true and is more likely to be not true for nonhedonic scales. If the equal interval assumption is violated, ordinal logistic regression models may be more appropriate because they do not require this assumption. A desirable attribute of ordinal logistic models is that one does not have to rerepresent ordinal categories of a response variable with numbers (such as 1–9).

The POM, proposed by McGullagh (1980), is a popular logistic regression model for ordinal categorical response because of its ease of interpretation. In this model, each observed ordinal response variable Y can be viewed as an underlying continuous, but unobservable, response variable Z (often called a latent variable). We cannot observe Z directly but, instead, implicitly dissect its range into J class intervals at unknown cut-points or thresholds $\alpha_1 < \alpha_2 < \ldots < \alpha_{J-1}$, so that Y takes the value Y_1 if Z is below α_1, the value Y_2 if Z is between α_1 and α_2, and the value Y_J if Z is above α_{J-1}. The relationship between Y and Z can be described by the following equation,

$$Y = \begin{cases} Y_1 & \text{for } Z \leq \alpha_1 \\ Y_2 & \text{for } \alpha_1 \leq Z \leq \alpha_2 \\ \vdots \\ Y_{J-1} & \text{for } \alpha_{J-2} \leq Z \leq \alpha_{J-1} \\ Y_J & \text{for } \alpha_{J-1} \leq Z \end{cases} \qquad (\text{Eq. 7.22})$$

For example, the observed overall liking Y on the 9-point hedonic scale can be described using the latent variable Z as

$$Y = \begin{cases} \text{dislike extremely} & \text{for } Z \leq \alpha_1 \\ \text{dislike very much} & \text{for } \alpha_1 < Z \leq \alpha_2 \\ \text{dislike moderately} & \text{for } \alpha_2 < Z \leq \alpha_3 \\ \text{dislike slightly} & \text{for } \alpha_3 < Z \leq \alpha_4 \\ \text{neither dislike nor like} & \text{for } \alpha_4 < Z \leq \alpha_5 \\ \text{like slightly} & \text{for } \alpha_5 < Z \leq \alpha_6 \\ \text{like moderately} & \text{for } \alpha_6 < Z \leq \alpha_7 \\ \text{like very much} & \text{for } \alpha_7 < Z \leq \alpha_8 \\ \text{like extremely} & \text{for } \alpha_8 < Z \end{cases} \qquad (\text{Eq. 7.23})$$

Because the latent variant Z is a continuous variable varying from $-\infty$ to $+\infty$, regression techniques can be used. For $Z_i = 1, 2, \ldots, n$ consumers, the linear relationship between Z and explanatory variables X can be expressed as

$$Z_i = X_i'\beta + \sigma\varepsilon_i = \eta_i + \sigma\varepsilon_i, \qquad (\text{Eq. 7.24})$$

where Z_i is the unobserved response or latent variable for consumer i, which corresponds to the observed response Y_i; X_i' is a vector of covariates for consumer i; β is a vector of regression coefficients; ε_i is the ith random error; σ is a constant scale factor; and

$$\eta_i = X_i'\beta. \tag{Eq. 7.25}$$

The cumulative probability distribution of Y_i is given by

$$\begin{aligned} P(Y_i \le j) &= P(Z_i \le \alpha_j); \\ &= P(\eta_i + \varepsilon_i \le \alpha_j); \\ &= P(\varepsilon_i \le \alpha_j - \eta_i) \quad (j = 1, 2, \ldots, J-1) \end{aligned} \tag{Eq. 7.26}$$

It is further assumed that the errors ε_i are independently distributed according to the following standard logistic distribution or probability density function

$$f(\varepsilon) = \frac{e^\varepsilon}{(1+e^\varepsilon)^2}. \tag{Eq. 7.27}$$

The probability density function is graphed in Figure 7.1 and resembles a normal distribution. The cumulative density function of the standard logistic distribution is given by

$$F(\varepsilon) = \frac{e^\varepsilon}{1+e^\varepsilon}, \tag{Eq. 7.28}$$

which is a logistic function, as the name of the logistic regression indicates.

Then, the probability that the response of the ith consumer will fall in the jth category or below (denoted by P_{ij}), given X_i, satisfies the equation

$$p_{ij} = P(Y_i \le j) = P(\varepsilon_i \le \alpha_j - X_i'\beta) = F(\alpha_j - X_i'\beta). \tag{Eq. 7.29}$$

By the inverse of the cumulative density function of the error term, we have the equation

$$g(p_{ij}) = F^{-1}(\gamma_{ij}) = \alpha_j - X_i'\beta, \tag{Eq. 7.30}$$

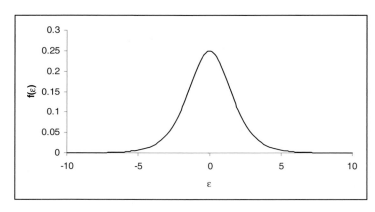

Figure 7.1. Probability density function for standard logistic distribution.

where $g(p_{ij})$ is called the link function in the generalized linear model. If the link function $g(p_{ij})$ takes the logit transformation, then the following proportional odds model is obtained

$$\text{logit}(p_{ij}) = \text{logit}[P(Y_i \leq j)] = \ln\left[\frac{P(Y \leq j)}{1 - P(Y \leq j)}\right]$$
$$= \alpha_j - X_i'\beta \quad (j = 1, 2, \ldots, J-1) \quad \text{(Eq. 7.31)}$$

This model can be rewritten as the following J equations

$$\text{logit}[p_{i1}] = \ln\left[\frac{p_{i1}}{1-p_{i1}}\right] = \alpha_1 - X_i'\beta;$$

$$\text{logit}[p_{i1} + p_{i2}] = \ln\left[\frac{p_{i1}+p_{i2}}{1-(p_{i1}+p_{i2})}\right] = \alpha_2 - X_i'\beta;$$

...

$$\text{logit}[p_{i1} + p_{i2} + \ldots + p_{i,J-1}] = \ln\left[\frac{p_{i1}+p_{i2}+\ldots+p_{i,J-1}}{1-(p_{i1}+p_{i2}+\ldots+p_{i,J-1})}\right] = \alpha_{J-1} - X_i'\beta. \quad \text{(Eq. 7.32)}$$

This model is also referred to as the cumulative logit model. It assumes proportional odds; that is, equal slope parameter β for all categories j ($j = 1, 2, \ldots, J-1$).

$$p_{i1} + p_{i2} + \ldots + p_{iJ} = 1.$$
$$\beta_1 = \beta_2 = \ldots = \beta_{J-1} = \beta. \quad \text{(Eq. 7.33)}$$

If $J = 2$, then proportional odds logit model is a binary logistic model. This model is overparameterized. Because the β vector typically includes a constant, say β_J, we have $J - 1$ regression equations, the intercepts of which are expressed in terms of J parameters. A solution is to eliminate the constant from β. Setting $\beta_J = 0$ in effect establishes the origin of the latent continuum Z; we already implicitly established the scale of Z by fixing the variance of the error to the variance of the standard logistic distribution ($\pi^2/3$). For convenience, the negative sign is absorbed into the slopes, rewriting the model as

$$\text{logit}[P(Y_i \leq j)] = \alpha_j + X_i'\beta \quad (j = 1, 2, \ldots, J-1) \quad \text{(Eq. 7.34)}$$

The remarkable thing about this model is that β coefficients do not depend on the category j and the position of the thresholds. That means that some of the αs may be close together, whereas others may be far apart, but the effects of the explanatory variables stay the same (Allison, 2001). The position of the αs does, of course, affect the intercepts and the relative numbers of cases that fall into the different categories. However, POM makes no assumption about the distance between observed categories.

When POM adequately represents the data, it is more parsimonious than the multinomial logit model, with POM having $(J + p - 2)$ independent parameters and the multinomial logit model having $p(J - 1)$ independent parameters.

With grouped data, the underlying continuous variable Z will have real existence, and the cut-points will usually be known (Rodriguez, 2005). For example, annual family income data are often collected in intervals (e.g., <$15,000, $15,001–$25,000, $25,001–$50,000, and >$50,000). In this case, the intervals are known, and we also know that the latent variable Z for the observed income Y is really continuous.

With ordinal categorical data, the underlying continuous variable will often represent a latent or unobservable trait, and the cut-points or thresholds will not be known. This would be the case for the degree of liking. For the POM model, there is no assumption about the equal distance between categories. The distances are not estimated by the POM model. They can be estimated by making appropriate assumptions about the distributions of the distances.

7.6.1. Example for POM

The data set for this example is the white corn tortilla chip data set described in Chapter 1. The example illustrates how to apply the proportional odds model to ordinal data using the 9-point hedonic scale (Meullenet et al., 2002). Overall liking with nine levels, from "dislike extremely" to "like extremely," is the response variable for the POM. The POM can be used for different purposes, including testing product differences, investigating demographic effects, and relating other product attributes to overall liking (i.e., similar to with preference mapping techniques).

7.6.2. Testing Product Differences

Sensory professionals are familiar with analysis of variance for significance testing. Similar to analysis of variance, POM can be used to assess the differences in the liking ratings for the 11 white corn tortilla chip products. Let us build a proportional odds model to describe the relationship between the response variable Overall (i.e., overall liking) and an explanatory variable Product. For the overall liking scores obtained using the 9-point hedonic scale, the probabilities cumulated from either "dislike extremely" to "like extremely" or from "like extremely" to "dislike extremely" can be modeled. It is usually more interesting to model the probabilities cumulated from "like extremely" to "dislike extremely", and the following cumulative probabilities can be defined:

$\theta_1 = \pi_1$, probability of "like extremely";

$\theta_2 = \pi_1 + \pi_2$, probability of "like very much" or "like extremely";

$\theta_3 = \pi_1 + \pi_2 + \pi_3$, probability of "like moderately," "like very much," or "like extremely";

. . .

$\theta_8 = \pi_1 + \pi_2 + \pi_3 + \pi_4 + \pi_5 + \pi_6 + \pi_7 + \pi_8$, probability of "dislike very much," "dislike moderately," "dislike slightly," "neither dislike nor like," "like slightly," "like moderately," "like very much," or "like extremely," (Eq. 7.35)

where π_1 = probability of "like extremely," π_2 = probability of "like very much," π_3 = probability of "like moderately," π_4 = probability of "like slightly," π_5 = probability of "neither dislike nor like," π_6 = probability of "dislike slightly," π_7 = probability of "dislike moderately," π_8 = probability of "dislike very much," π_9 = probability of "dislike extremely," and $\pi_1 + \pi_2 + \pi_3 + \pi_4 + \pi_5 + \pi_6 + \pi_7 + \pi_8 + \pi_9 = 1$.

Then the corresponding cumulative logits (log odds ratio) can be constructed:

$$\text{logit}(\theta_1) = \text{logit}\left(\frac{\theta_1}{1-\theta_1}\right) = \text{logit}\left(\frac{\pi_1}{\pi_2+\pi_3+\pi_4+\pi_5+\pi_6+\pi_7+\pi_8+\pi_9}\right);$$

$$\text{logit}(\theta_2) = \text{logit}\left(\frac{\theta_2}{1-\theta_2}\right) = \text{logit}\left(\frac{\pi_1+\pi_2}{\pi_3+\pi_4+\pi_5+\pi_6+\pi_7+\pi_8+\pi_9}\right);$$

$$\text{logit}(\theta_3) = \text{logit}\left(\frac{\theta_3}{1-\theta_3}\right) = \text{logit}\left(\frac{\pi_1+\pi_2+\pi_3}{\pi_4+\pi_5+\pi_6+\pi_7+\pi_8+\pi_9}\right);$$

$$\text{logit}(\theta_4) = \text{logit}\left(\frac{\theta_4}{1-\theta_4}\right) = \text{logit}\left(\frac{\pi_1+\pi_2+\pi_3+\pi_4}{\pi_5+\pi_6+\pi_7+\pi_8+\pi_9}\right);$$

$$\text{logit}(\theta_5) = \text{logit}\left(\frac{\theta_5}{1-\theta_5}\right) = \text{logit}\left(\frac{\pi_1+\pi_2+\pi_3+\pi_4+\pi_5}{\pi_6+\pi_7+\pi_8+\pi_9}\right);$$

$$\text{logit}(\theta_6) = \text{logit}\left(\frac{\theta_6}{1-\theta_6}\right) = \text{logit}\left(\frac{\pi_1+\pi_2+\pi_3+\pi_4+\pi_5+\pi_6}{\pi_7+\pi_8+\pi_9};\right)$$

$$\text{logit}(\theta_7) = \text{logit}\left(\frac{\theta_7}{1-\theta_7}\right) = \text{logit}\left(\frac{\pi_1+\pi_2+\pi_3+\pi_4+\pi_5+\pi_6+\pi_7}{\pi_8+\pi_9}\right);$$

$$\text{logit}(\theta_8) = \text{logit}\left(\frac{\theta_8}{1-\theta_8}\right) = \text{logit}\left(\frac{\pi_1+\pi_2+\pi_3+\pi_4+\pi_5+\pi_6+\pi_7+\pi_8}{\pi_9}\right). \quad \text{(Eq. 7.36)}$$

The proportional odds model is then given below

$$\text{logit}(\theta_j) = \text{logit}\left(\frac{\theta_j}{1-\theta_j}\right) = \alpha_j + X'\beta \quad (j=1, 2, \ldots, 8). \quad \text{(Eq. 7.37)}$$

Because Product is categorical, dummy variables are used to represent Product in the model. If the last product (i.e., TOR) is used as a reference, then the regression coefficient for TOR is zero ($\beta_{TOR} = 0$). The POM becomes

$$\begin{aligned}\text{logit}(\theta_j) = {} & \alpha_j + \beta_{BWY}X_{BWY} + \beta_{GMG}X_{GMG} + \beta_{GUY}X_{GUY} + \beta_{MED}X_{MED} + \\ & \beta_{MIS}X_{MIS} + \beta_{MIT}X_{MIT} + \beta_{OAK}X_{OAK} + \beta_{SAN}X_{SAN} + \beta_{TOB}X_{TOB} + \\ & \beta_{TOM}X_{TOM} + \beta_{TOR}X_{TOR} \quad (j=1, 2, \ldots, 8),\end{aligned} \quad \text{(Eq. 7.38)}$$

where the explanatory variables, X, are dummy variables taking values of 1 or 0, and the sum of X is 1, βs are regression coefficients independent of category j, and subscripts are the names of products. For any category $j = 1, 2, \ldots, 8$, the logit between any two products (e.g., BWY and GMG) is

$$\text{logit}(\theta_{jBWY}) - \text{logit}(\theta_{jGMG}) = (\alpha_j + \beta_{BWY}) - (\alpha_j + \beta_{GMG}) = \beta_{BWY} - \beta_{GMG}; \quad \text{(Eq. 7.39)}$$

therefore the odds ratio is

$$\frac{\theta_{jBWY}}{\theta_{jGMG}} = e^{\beta_{BWY} - \beta_{GMG}}, \quad \text{(Eq. 7.40)}$$

which is proportional to ($\beta_{BWY} - \beta_{GMG}$) and independent of category j on the scale.

The proportional odds model is fitted using the following SAS statements. The DESCENDING option is used to sort the response in decreasing order (i.e., "like extremely" to "dislike extremely") because we are interested in liking. It is a good habit to check the "Response Profiles" table to verify that response levels are appropriately ordered.

```
PROC LOGISTIC DATA=Chips DESCENDING;
      CLASS Product/PARAM=GLM;
      MODEL  Overall=Product/ LINK=Logit  RSQUARE  SCALE=None
AGGREGATE;
      TITLE1 'POM: Overall=Product';
      CONTRAST "BWY vs GMG" Product 1 -1 0 0 0 0 0 0 0 0/ESTIMATE;
      CONTRAST "BWY vs GUY" Product 1 0 -1 0 0 0 0 0 0 0/ESTIMATE;
      CONTRAST "BWY vs MED" Product 1 0 0 -1 0 0 0 0 0 0/ESTIMATE;
      CONTRAST "BWY vs MIS" Product 1 0 0 0 -1 0 0 0 0 0/ESTIMATE;
      CONTRAST "BWY vs MIT" Product 1 0 0 0 0 -1 0 0 0 0/ESTIMATE;
      CONTRAST "BWY vs OAK" Product 1 0 0 0 0 0 -1 0 0 0/ESTIMATE;
      CONTRAST "BWY vs SAN" Product 1 0 0 0 0 0 0 -1 0 0/ESTIMATE;
      CONTRAST "BWY vs TOB" Product 1 0 0 0 0 0 0 0 -1 0/ESTIMATE;
      CONTRAST "BWY vs TOM" Product 1 0 0 0 0 0 0 0 0 -1 0/ESTIMATE;
      CONTRAST "BWY vs TOR" Product 1 0 0 0 0 0 0 0 0 -1/ESTIMATE;
RUN;
```

```
           Score Test for the Proportional Odds Assumption
                   Chi-Square   DF   Pr > ChiSq
                     71.8628    70     0.4159

       Deviance and Pearson Goodness-of-Fit Statistics
          Criterion  Value    DF   Value/DF   Pr > ChiSq
          Deviance   78.1604  70    1.1166      0.2356
          Pearson    72.3423  70    1.0335      0.4005

              Testing Global Null Hypothesis: BETA = 0
              Test           Chi-Square   DF     Pr > ChiSq
              Likelihood Ratio              124.3366   10   <.0001
              Score          117.3428     10     <.0001
              Wald           118.7041     10     <.0001

                  Type 3 Analysis of Effects Wald
              Effect    DF  Chi-Square  Pr > ChiSq
              Product   10    118.7041    <.0001

         Analysis of Maximum Likelihood Estimates Standard
                          Wald Odds
Parameter           DF  Estimate  Error    Chi-       Pr >      Ratio
                                           Square     ChiSq
Intercept H9-LX    1   -1.9351   0.2285   71.6962   <.0001
Intercept H8-LV    1   -0.0160   0.2041    0.0062    0.9373
Intercept H7-LM    1    1.2235   0.2090   34.2852   <.0001
Intercept H6-LS    1    1.9710   0.2147   84.2778   <.0001
```

```
Intercept H5-DL   1    2.4212   0.2194   121.8188   <.0001
Intercept H4-DS   1    3.3851   0.2357   206.1998   <.0001
Intercept H3-DM   1    4.1352   0.2605   251.9249   <.0001
Intercept H2-DV   1    4.9991   0.3149   251.9521   <.0001

Product BYW       1   -0.9427   0.2840    11.0170   0.0009   0.390
Product GMG       1   -1.7550   0.2863    37.5825   <.0001   0.173
Product GUY       1   -1.3918   0.2850    23.8544   <.0001   0.249
Product MED       1   -1.9294   0.2871    45.1602   <.0001   0.145
Product MIS       1   -0.9776   0.2841    11.8441   0.0006   0.376
Product MIT       1   -0.5580   0.2838     3.8652   0.0493   0.572
Product OAK       1   -1.6840   0.2868    34.4712   <.0001   0.186
Product SAN       1   -0.1640   0.2843     0.3326   0.5641   0.849
Product TOB       1    0.1830   0.2855     0.4110   0.5215   1.201
Product TOM       1   -0.9032   0.2840    10.1170   0.0015   0.405
Product TOR       0    0                    .        .        .
```

```
              Contrast Test Results     Wald
   Contrast       DF   Chi-Square   Pr > ChiSq
   BWY vs GMG     1     8.4199       0.0037
   BWY vs GUY     1     2.5812       0.1081
   BWY vs MED     1    12.3813       0.0004
   BWY vs MIS     1     0.0156       0.9007
   BWY vs MIT     1     1.8740       0.1710
   BWY vs OAK     1     6.9765       0.0083
   BWY vs SAN     1     7.5741       0.0059
   BWY vs TOB     1    15.5583       <.0001
   BWY vs TOM     1     0.0199       0.8879
   BWY vs TOR     1    11.0170       0.0009
```

The first thing of interest is the goodness of fit of the specified model. In this case, the Deviance and Pearson Goodness-of-Fit Statistics Table needs to be examined. From the table, the probability of deviance is $p = 0.2356$, indicating that the model fits the data at $\alpha = 0.05$. The proportional odds assumption needs to be checked, as it is crucial for the proportional odds model. If the assumption is violated, the results from the POM are no longer valid (Agresti, 2003; Allison, 2001; Bender and Grouven, 1997, 1998; Lawal, 2004). Because the score test from the SAS output clearly indicates that the proportional odds assumption is not rejected ($p = 0.4159 > \alpha = 0.05$) in this example, the model can be used to test for Product significance. The Type 3 test indicates that Product is highly significant ($p < 0.0001$); that is, there are significant differences among the products, but this alone does not tell which products are different. The Wald χ^2 statistic (i.e., the squared z or t statistic) is usually used to look at the relative differences between the products by comparing regression coefficients. PROC LOGISTIC provides a convenient way through the CONTRAST and ESTIMATE statements to perform pairwise comparisons of products. The Contrast Test Results in the SAS output show that BWY (Best Yet) is significantly different from GMG (Green Mountain Gringo), MED (Medallion), OAK (Oak Creek Farms), SAN (Santita's), TOB (Tostito's bite size), and TOR (Tostito's restaurant) but not significantly different from the remaining products (GUY [Guy's restaurant], MIS [Mission strips], MIT [Mission triangle], and TOM [Tom's]). Similarly, we could perform all possible pairwise comparisons of the 11 products, and the results are given in Table 7.6. Because the nine categories of the hedonic scale are usually assumed to have

Table 7.6. GLM and POM mean separation results.

Products	Mean of Overall Liking Scores (GLM with LSD test)	Regression Coefficients (POM)
TOB	7.25a	0.183a
TOR	7.15ab	0.000a
SAN	7.09ab	−0.164ab
MIT	6.65bc	−0.558bc
BYW	6.39cd	−0.9427cd
TOM	6.29cd	−0.9032cd
MIS	6.26cd	−0.9776cd
GUY	5.89de	−1.3918de
OAK	5.51ef	−1.684e
GMG	5.46ef	−1.755e
MED	5.28f	−1.9294e

BWY, Best Yet; GMG, Green Mountain Gringo; MED, Medallion; OAK, Oak Creek Farms; SAN, Santita's; TOB, Tostito's bite size; TOR, Tostito's restaurant; GUY, Guy's restaurant; MIS, Mission strips; MIT, Mission triangle; TOM, Tom's.
Note: Means within a column with different letters are significantly different from each other at the 95% confidence level.

equal intervals and are represented using numbers 1–9 in consumer studies, sensory professionals often use a regular GLM model for the significance test. The results from the GLM model are also given in Table 7.6. The table shows that the POM found significant difference between TOR and MIT, whereas GLM did not. In contrast, GLM reported a significant difference between GUY and MED, whereas POM did not. Overall, the GLM and POM give similar results, which is evidence that the intervals between categories on the hedonic scale may be equal or approximately equal. If the GLM and POM give very different results, the equal intervals may not be true, and the results from the GLM may be questionable. In addition, the interpretation of the POM in terms of odds is more meaningful. For example, the odds ratio of 1.201 ($e^{0.183}$) for TOB versus TOR (Table 7.6) means that the odds of consumers rating product TOB as liked was 1.201 times the odds for product TOR. This means that consumers tended to rate product TOB higher in overall liking than they did product TOR.

7.6.3. Testing Demographic Information Effects

POM can also be used to examine the effects of demographic information on overall liking if demographic information is available. In the case of tortilla chips, demographic information such as age group, gender, preferred shape, consumption, and preferred brand were available, and their effects on overall acceptance were examined. The SAS code for the POM, including demographic information effects, is provided below. In the SAS code, the SELECTION=backward (please note that forward and stepwise are also available) option with SLS = 0.05 and SLE = 0.10 is used for stepwise regression to select significant variables.

The SAS output shows that the POM fits the data (probability of deviance $p \approx 1.0 > \alpha = 0.05$) and that the proportional odds assumption is not rejected ($p = 0.4511 > \alpha = 0.05$). By backward elimination, the age group and preferred brand were significant, but other variables were not. The positive coefficient (0.0249) for age group means that for a 1-unit increase in age group, the expected change in log of odds of liking tortilla chips is 0.0249, indicating that older consumers tended to give slightly higher scores than younger consumers. The negative coefficient (−0.5374) for preferred brand 0 (i.e., no preferred brand) means the odds ratio was 0.5843 ($e^{-0.5374}$) for preferred brand (1) versus preferred brand (0), indicating that preferred brand chips were liked more than nonpreferred brand chips and that consumers somehow recognized their preferred brand chips.

```
*Fit the POM;
PROC LOGISTIC DATA=Chips DESCENDING;
    CLASS Product Gender Consumptions PreferredShape Pre-
ferredBrand /PARAM=GLM;
    MODEL Overall=Product Gender PreferredShape AgeGroup
Consumptions      PreferredBrand       /SELECTION=Backward
SLS=0.05 SLE=0.10 LINK=Logit RSQUARE LACKFIT
    SCALE=None AGGREGATE;
    TITLE1 'POM: Overall acceptance=Product & Demographical
Variables';
RUN;
```

```
              Response Profile
         Ordered            Total
         Value     Overall  Frequency
         1         H9-LX    58
         2         H8-LV    206
         3         H7-LM    234
         4         H6-LS    136
         5         H5-DL    67
         6         H4-DS    97
         7         H3-DM    40
         8         H2-DV    23
         9         H1-DX    18
```

Probabilities modeled are cumulated over the lower Ordered Values.

```
Score Test for the Proportional Odds Assumption
     Chi-Square   DF   Pr > ChiSq
      84.9305     84     0.4511

       Model Fit Statistics   Intercept
     Intercept and
     Criterion      Only       Covariates
     AIC            3383.356   3272.496
     SC             3421.586   3368.072
     -2 Log L       3367.356   3232.496

R-Square 0.1422    Max-rescaled R-Square 0.1454

   Testing Global Null Hypothesis: BETA = 0
   Test                Chi-Square   DF   Pr > ChiSq
   Likelihood Ratio    134.8596     12    <.0001
   Score               127.7299     12    <.0001
   Wald                127.0209     12    <.0001
```

Summary of Backward Elimination

Step	Effect Removed	DF	Number In	Wald Chi-Square	Pr > ChiSq
1	Gender	1	4	0.0004	0.9831
2	Consumptions	4	3	7.2138	0.1250

Deviance and Pearson Goodness-of-Fit Statistics

Criterion	Value	DF	Value/DF	Pr > ChiSq
Deviance	2032.9025	3284	0.6190	1.0000
Pearson	3247.5235	3284	0.9889	0.6713

Number of unique profiles: 413

Type 3 Analysis of Effects

Effect	DF	Wald Chi-Square	Pr > ChiSq
ProductID	10	66.2572	<.0001
AgeGroup	1	4.5034	0.0338
PreferredBrand	1	6.9030	0.0086

Analysis of Maximum Likelihood Estimates

Parameter		DF	Estimate	Standard Error	Wald Chi-Square	Pr > ChiSq
Intercept	H9-LX	1	-2.5672	0.3927	42.7369	<.0001
Intercept	H8-LV	1	-0.6375	0.3764	2.8685	0.0903
Intercept	H7-LM	1	0.6138	0.3769	2.6517	0.1034
Intercept	H6-LS	1	1.3663	0.3789	13.0020	0.0003
Intercept	H5-DL	1	1.8186	0.3809	22.7927	<.0001
Intercept	H4-DS	1	2.7870	0.3895	51.1949	<.0001
Intercept	H3-DM	1	3.5401	0.4046	76.5647	<.0001
Intercept	H2-DV	1	4.4048	0.4414	99.6000	<.0001
ProductID	BYW	1	-0.4731	0.3356	1.9871	0.1586
ProductID	GMG	1	-1.2787	0.3365	14.4372	0.0001
ProductID	GUY	1	-1.0007	0.3220	9.6565	0.0019
ProductID	MED	1	-1.4674	0.3371	18.9487	<.0001
ProductID	MIS	1	-0.7329	0.3013	5.9179	0.0150
ProductID	MIT	1	-0.3272	0.3006	1.1847	0.2764
ProductID	OAK	1	-1.2077	0.3371	12.8364	0.0003
ProductID	SAN	1	0.2439	0.3255	0.5614	0.4537
ProductID	TOB	1	0.1827	0.2856	0.4093	0.5223
ProductID	TOM	1	-0.4273	0.3356	1.6207	0.2030
ProductID	TOR	0	0	.	.	.
AgeGroup		1	0.0249	0.0117	4.5034	0.0338
PreferredBrand 0		1	-0.5374	0.2045	6.9030	0.0086
PreferredBrand 1		0	0	.	.	.

7.6.4. Internal Preference Map

Consumer preference of tortilla chips can be studied using the external and internal preference modeling techniques that have been employed for many foods (Arditti, 1997; Meullenet et al., 2000; 2002). Internal preference modeling uses only the consumer data to determine consumer preference patterns, whereas external preference modeling relates consumer preference data to descriptive sensory information or instrumental data (Lawlor and Delahunty, 2000). By internal preference mapping, we do not refer to MDPREF but, rather, to simply modeling overall liking from other attributes evaluated on the products by consumers. POM can also be used for both internal and external preference maps if it fits the data (Meullenet et al., 2002).

Because consumer variables (such as acceptance of appearance, flavor, and texture) from consumer testing are correlated, principal components (continuous variables) are often used instead of the consumer variables for preference mapping. When continuous variables are used in POM, however, the conventional goodness-of-fit statistics provided in SAS procedures may be no longer valid if there are too many unique data patterns or profiles. The goodness-of-fit testing for ordinal response regression models using continuous variables is not available from the modern statistical packages. This is one of the disadvantages of POM for applications in sensory science. Fortunately, there are two proposed methods available for assessing goodness of fit for ordinal response models, using continuous variables (Lipsitz et al., 1996; Pulkstenis and Robinson, 2004). SAS macros for the proposed methods are available from http://lib.stat.cmu.edu/general/.

To apply POM to internal consumer data, it is assumed that the intervals between the nine categories (from "dislike extremely" to "like extremely") for acceptance scores of appearance, flavor, and texture are equal and that the categories are represented by numbers from 1 to 9, respectively. Because of collinearity between acceptance scores of appearance, flavor, and texture, principal components are used as covariates in POM. According to the backward elimination method, the "unimportant" sensory variables are manually eliminated, and only important sensory variables are used to compute principal components for POM. The final SAS codes are given below.

```
*Compute principal components (Flav=Flavor, Text=Texture, App_Flav=Appearance*Flavor);
PROC PRINCOMP DATA=Chips OUT=PCsIn OUTSTAT=StatIn;
    VAR Flav Text App_Flav;
RUN;
*Fit POM using principal components;
PROC LOGISTIC DATA=PCsIn DESCENDING OUT=ParamsIn;
    MODEL Overall=PRIN1-PRIN3 /SELECTION=Backward SLS=0.05 SLE=0.10 LINK=Logit RSQUARE LACKFIT SCALE=None AGGREGATE;
    TITLE1 'Internal Preference Modeling For Overall Acceptance';
    OUTPUT OUT=PredOverallIn(KEEP=Product _LEVEL_ Pred) P=Pred;
RUN;
*Perform GOF testing (OL=Overall Like);
%Include "C:\GOF_ContinuousVariables.sas";
%goodness(vars= , contvar=PRIN1 PRIN2 PRIN3, k=0, dset=PCsIn, y=OL ,j=9, alpha=0.05);
```

The POM model for consumer overall acceptance using the principal components as covariates is presented in Table 7.7. The probability of the goodness-of-fit statistic achieved by Lipsitz's (1996) method is 0.021, indicating that the model does not fit the data at $\alpha = 0.05$ but fits the data at $\alpha = 0.01$. We use here $\alpha = 0.01$ for purposes of illustrating the application of POM. If the POM model fits the data, then the proportional odds assumption is checked. For this example, the proportional odds assumption is satisfied ($p = 0.35$ for χ^2 of 15.42 with DF = 14). Two principal components (PC_1 and PC_2)

Table 7.7. Parameter estimates and statistics for proportional odds model for internal preference mapping.

Parameter[1]	DF	Estimate	Standard Error	Chi-Square	p > Chi-Square	Odds Ratio
α_1	1	−6.22	0.24	658.96	<.0001	
α_2	1	−2.75	0.15	347.59	<.0001	
α_3	1	0.42	0.12	12.39	0.0004	
α_4	1	2.53	0.15	277.73	<.0001	
α_5	1	3.88	0.19	435.41	<.0001	
α_6	1	6.28	0.26	576.95	<.0001	
α_7	1	7.84	0.32	594.36	<.0001	
α_8	1	9.40	0.40	544.35	<.0001	
PC_1	1	2.56	0.10	719.52	<.0001	12.96
PC_2	1	−0.76	0.10	54.85	<.0001	0.47

[1]PC = principal component.
Note: PC_i means the *i*th principal component, PC_1 = 0.60Flavor + 0.53Texture + 0.60Appearance × Flavor; PC_2 = −0.40Flavor + 0.85Texture − 0.35Appearance × Flavor.

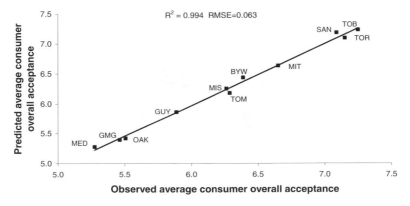

Figure 7.2. Predicted versus observed average consumer overall acceptance. Predicted average scores were obtained from consumer appearance, flavor, and texture acceptance, using a proportional odds model.

are statistically significant. From the principal components analysis, PC_1 and PC_2 can be expressed as 0.60Flavor + 0.53Texture + 0.60Appearance × Flavor and −0.40Flavor + 0.85Texture − 0.35Appearance × Flavor, respectively. Keep in mind that the important sensory variables (i.e., flavor, texture, and appearance by flavor) associated with the principal components are selected using the backward elimination method. The parameter estimates for PC_1 and PC_2 are 2.56 and −0.76, respectively. To examine the effects of the original sensory variables on consumer overall acceptance, $2.56PC_1 - 0.76PC_2$ was reexpressed as 1.83Flavor + 0.72Texture + 1.81Appearance × Flavor. The regression coefficients in this expression clearly show that the most influential sensory variable was flavor (1.83), followed by the interaction (1.81) of appearance with flavor and texture (0.72). The positive coefficients indicated that consumer overall acceptance had a positive association with the sensory variables of flavor, appearance, and texture. On the basis of the predicted mean probabilities, the predicted mean scores of consumer overall acceptance were calculated to be compared with the observed mean scores for the 11 tortilla chips (Figure 7.2). The corresponding R^2_{Score} and $RMSE_{Score}$ values were 0.99 and 0.06 (Figure 7.2), respectively, which also indicates that the model predicted the mean scores well for consumer overall acceptance. The R^2_{score} and $RMSE_{score}$ are useful for comparison of the proportional odds model with other regression models in terms of fitting the mean scores of the response. For example, R^2_{score} and $RMSE_{score}$ can be used to assess the goodness of fit of the proportional odds model for the mean scores against the goodness of fit of the partial least squares regression model that is a popular tool for preference modeling (Meullenet et al., 2000).

7.6.5. External Preference Map

Similarly, POM can be used for external preference mapping. Descriptive intensity means that visual appearance, flavor, and texture attributes are used for external preference mapping. Because the descriptive variables are correlated, principal components are calculated as covariates in POM. The SAS codes to perform this analysis are given here.

```
*Compute principal components (Aftertaste=Aftertaste salt,
a=color a);
PROC PRINCOMP DATA=Chips OUT=PCsEx OUTSTAT=StatEx;
    VAR AfterSalt Crispness a;
RUN;

*POM for external preference map;
PROC LOGISTIC DATA=PCsEx DESCENDING OUT=ParamsEx;
    CLASS Shape_App/PARAM=GLM;
    MODEL Overall=PRIN1-PRIN3 Shape_App/SELECTION=Backward
SLS=0.05  SLE=0.10  LINK=Logit  RSQUARE  LACKFIT  SCALE=None
AGGREGATE;
    TITLE1 'External Preference Modeling For Overall
Acceptance';
    OUTPUT OUT=PredOverallEx(KEEP=Product _LEVEL_ Pred)
P=Pred;
RUN;

*Perform GOF testing (OL=Overall Like);
%Include "C:\GOF_ContinuousVariables.sas";
%goodness(vars= , contvar=PRIN1 PRIN2 PRIN3, k=0, dset=PCsIn,
y=OL ,j=9, alpha=0.05);
```

The goodness-of-fit test (Lipsitz's method) shows that the POM adequately fits the data ($p = 0.7 > \alpha = 0.05$) for overall acceptance. The proportional odds assumption is also met ($\chi^2 = 9.25$, DF = 14, $p = 0.81$). The parameter estimates are listed in Table 7.8. Two (PC_1 and PC_2) of three principal components are selected as important covariates, which is equivalent to one flavor descriptive attribute (aftertaste salt), one appearance attribute (instrumental color a), and one texture descriptive attribute (crispness) being significant contributors to the consumer overall acceptance of tortilla chips (Table 7.8). Appearance, flavor, and texture descriptive attributes are included in the external preference map, confirming the results from the previously described internal preference map. It should not be surprising that aftertaste salt is the most important flavor attribute to acceptance because American consumers like salt. In addition, crispness is one of the most important texture attributes for all chips, including tortilla chips. As expected, crispness increases

Table 7.8. Parameter estimates for the proportional odds model for external preference modeling.

Parameter[1]	DF	Overall Acceptance			Appearance Acceptance			Flavor Acceptance			Texture Acceptance			Purchase Intent		
		Estimate	Chi-Square	P > Chi-Square	Estimate	Chi-Square	P > Chi-Square	Estimate	Chi-Square	P > Chi-Square	Estimate	Chi-Square	P > Chi-Square	Estimate	Chi-Square	P > Chi-Square
α_1	1	−2.84	415.84	<.0001	−2.12	267.33	<.0001	−2.70	429.79	<.0001	−2.45	415.11	<.0001	−2.01	388.08	<.0001
α_2	1	−0.93	147.08	<.0001	−0.38	16.02	<.0001	−0.94	147.42	<.0001	−0.61	69.85	<.0001	−0.40	31.108	<.0001
α_3	1	0.30	17.84	<.0001	0.74	58.05	<.0001	0.01	0.01	0.9248	0.47	42.89	<.0001	0.83	117.76	<.0001
α_4	1	1.04	176.28	<.0001	1.40	177.84	<.0001	0.80	110.99	<.0001	1.19	215.12	<.0001	2.26	413.11	<.0001
α_5	1	1.49	290.76	<.0001	1.90	271.84	<.0001	1.26	230.71	<.0001	1.71	336.34	<.0001			
α_6	1	2.46	419.67	<.0001	3.21	358.93	<.0001	2.20	407.53	<.0001	2.90	409.47	<.0001			
α_7	1	3.21	390.25	<.0001	4.23	272.92	<.0001	2.92	421.12	<.0001	3.72	329.43	<.0001			
α_8	1	4.07	289.04	<.0001	5.98	105.13	<.0001	3.96	315.98	<.0001	5.37	142.33	<.0001			
PC_{1o}	1	0.50	106.58	<.0001												
PC_{2o}	1	0.20	9.42	0.0021												
PC_{2a}	1				−0.17	6.79	0.0092									
PC_{3a}	1				−0.25	11.73	0.0006									
PC_{4a}	1				−0.85	14.90	0.0001									
Shape: Round	1				−0.30	2.95	0.0860									
Shape: Strip	1				−0.73	19.52	<.0001									
Shape: Triangle	0				0.00	...	...									

PC_{1f}	1	0.35	95.25	<.0001
PC_{3f}	1	0.22	10.53	0.0012
PC_{5f}	1	0.35	6.96	0.0083
PC_{6f}	1	−1.68	35.44	<.0001
PC_{1t}	1	0.26	40.39	<.0001
PC_{2t}	1	−0.11	5.88	0.0154
PC_{3t}	1	0.47	72.71	<.0001
PC_{1p}	1	0.55	125.85	<.0001
PC_{2p}	1	0.28	17.90	<.0001

^{1}PC = principal component, PC_i = ith principal component, subscripts (o, a, f, t, and p) stand for acceptance of overall, appearance, flavor and texture, and purchase intent, respectively.

Note: Models for each attribute are as follows:

Overall acceptance: PC_{1o} = 0.51SaltAftertaste + 0.67Crispness − 0.55a*, PC_{2o} = 0.75SaltAftertaste − 0.03Crispness + 0.66a*, **0.5PC_{1o} + 0.2PC_{2o} = 0.41SaltAftertaste + 0.32Crispness − 0.14a*.**

Appearance acceptance: PC_{2a} = 0.18ToastedCorn + 0.46GrainFlecks − 0.32CharMarks + 0.75CharMarks + 0.13a*, PC_{4a} = −0.52DegreeOfWhiteness + 0.77GrainFlecks + 0.34CharMarks + 0.17a*, **0.85PC_{2a} + 0.78PC_{3a} + 0.43PC_{4a} = 0.24DegreeOfWhiteness − 0.68GrainFlecks − 0.42CharMarks − 0.33a*.**

Flavor acceptance: PC_{1f} = 0.46ToastedCorn + 0.49SaltAftertaste − 0.45ToastedGrainAftertaste − 0.24SweetAftertaste − 0.39Cardboard + 0.36GrainComplex, PC_{3f} = −0.16ToastedCorn + 0.44SaltAftertaste − 0.40ToastedGrainAftertaste + 0.69SweetAftertaste + 0.06Cardboard − 0.37GrainComplex, PC_{5f} = −0.30ToastedCorn + 0.74SaltAftertaste − 0.53ToastedGrainAftertaste − 0.26SweetAftertaste + 0.09Cardboard − 0.03 GrainComplex, PC_{6f} = −0.48ToastedCorn + 0.02SaltAftertaste − 0.41ToastedGrainAftertaste − 0.10SweetAftertaste + 0.51Cardboard + 0.57GrainComplex, **0.35PC_{1f} + 0.22PC_{3f} + 0.35PC_{5f} − 1.68PC_{6f} = 0.84ToastedCorn + 0.50SaltAftertaste + 0.63ToastedGrainAftertaste + 0.14SweetAftertaste − 0.95Cardboard − 0.93GrainComplex.**

Texture acceptance: PC_{1t} = −0.34Hardness + 0.17Toothpack + 0.51Crispness + 0.37OilyFilm − 0.43MoistAbsorption + 0.51PersistOfCrisp + 0.11LooseParticles, PC_{2t} = 0.43Hardness + 0.56Toothpack − 0.32Crispness − 0.08OilyFilm − 0.29MoistAbsorption + 0.12PersistOfCrisp + 0.55LooseParticles, PC_{3t} = −0.45Hardness + 0.33Toothpack + 0.33Crispness − 0.54OilyFilm + 0.40MoistAbsorption − 0.10PersistOfCrisp + 0.35LooseParticles, **0.26PC_{1t} − 0.11PC_{2t} + 0.47PC_{3t} = −0.35Hardness + 0.14Toothpack + 0.32Crispness − 0.15OilyFilm + 0.11MoistAbsorption + 0.08PersistOfCrisp + 0.13LooseParticles.**

Purchase intent: PC_{1p} = 0.51SaltAftertaste + 0.67Crispness − 0.55a*, PC_{2p} = 0.75SaltAftertaste − 0.03Crispness + 0.66a*, **0.55PC_{1p} + 0.28PC_{2p} = 0.49SaltAftertaste + 0.36Crispness − 0.12a*.**

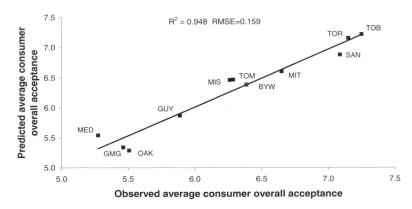

Figure 7.3. Predicted versus observed average consumer overall acceptance. Predicted average scores were obtained from sensory descriptive attributes, using a proportional odds model.

consumer overall acceptance, but its contribution to overall acceptance is lower than that of aftertaste salt. Instrumental color a negatively contributes to consumer overall acceptance. The R^2_{Score} and $RMSE_{Score}$ values for the average scores of consumer overall acceptance are 0.95 and 0.16 (Figure 7.3), respectively, indicating that the model could accurately predict the mean scores of overall acceptance.

7.7. Conclusions

In this chapter, we hopefully demonstrated that proportional odds models offer sensory scientists an alternative choice for internal/external preference modeling. The use of principal components in a proportional odds model can be used to combat collinearity resulting from correlation between sensory variables. Proportional odds models can model the structure of ordinal categorical responses and estimate the mean scores of the responses.

References

Achen, C.H. 1982. Interpreting and using regression. Sage University Paper.
Agresti, A. 2002. Categorical data analysis, 2nd edition New York: Wiley.
Agresti, A. 2003. Dealing with discreteness: Making "exact" confidence intervals for proportions, differences of proportions, and odds ratios more exact. Stat. Methods Med. Res. 12:3–21.
Allison, P.D. 2001. Logistic regression using the SAS system: theory and application. New York: Wiley.
Arditti, S. 1997. Preference mapping: A case study. Food Q. Pref. 8(5):323–327.
Hosmer, D.W. and S. Lemeshow. 2000. Applied logistic regression. Hoboken, NJ: Wiley.
Jennings, D.E. 1986. Judging inference adequacy in logistic regression. J. Am. Stat. Assoc. 81:471–476.
Lawlor, J.B. and C.M. Delahunty. 2000. The sensory profile and consumer preference for ten specialty cheeses. Int. J. Dairy Technol. 53:28–36.
Lawal, H.B. 2004. Review of non-independence, asymmetry, shew-symmetry and point-symmetry models in the analysis of social mobility data. Quant. Q. 38:259–289.
Lipsitz, S.R., G.M. Fitzmaurice, and G. Molenberghs. 1996. Goodness-of-fit tests for ordinal response regression models. Appl. Stat. 45(2):175–190.
McGullagh, P. 1980. Regression models for ordinal data (with discussion). J. R. Stat. Soc. Ser. B 42:109–142.
Meullenet, J.-F., V.K. Griffin, K. Carson, G. Davis, S. Davis, J. Gross, J.A. Hankins, E. Sailer, C. Sitakalin, S. Suwansri, and A.L. Vasquez Caicedo. 2000. External rice preference mapping for Asian consumers living in the United States. J. Sensory Stud. 16:73–93.

Meullenet, J.-F., V.K. Griffin, K. Carson, G. Davis, S. Davis, J. Gross, J.A. Hankins, E. Sailer, C. Sitakalin, S. Suwansri, and A.L. Vasquez Caicedo. 2001. Rice external preference mapping for Asian consumers living in the United States. J. Sensory Stud. 16:73–94.

Meullenet, J.-F., R. Xiong, M. Monsoor, T. Bellman-Horner, S. Zivanovic, P. Dias, H. Fromm, and Z. Liu. 2002. Preference mapping of commercial toasted white corn tortilla chips. J. Food Sci.

Nelder, J.A. and R.W.M. Wedderburn. 1972. Generalized linear models. J. R. Stat. Soc. A 135:370–384.

Pearson, K. 1900. On the criterion that a given system of deviations from the probable in the case of a correlated system of variables is such that it can reasonably be supposed to have arisen from random sampling. Philos. Mag. 50:157–176.

Pulkstenis, E. and T.J. Robinson. 2004. Goodness-of-fit tests for ordinal response regression models. Stat. Med. 23:999–1014.

Rud, O.P. 2001. Data mining cookbook: Modeling data for marketing, risk, and customer relationship management. New York: Wiley.

Sangketkit, C., G.P. Savage, R.J. Martin, B.P. Searle, and S.L. Mason. 2000. Sensory evaluation of new lines of oca (*Oxalis tuberosa*) grown in New Zealand. Food Q. Pref. 11:189–199.

Tepper, B.J., S.C. Young, and R.M. Nayga. 1997. Understanding food choice in adult men: Influence of nutrition knowledge, food beliefs and dietary restraint. Food Q. Pref. 8(4):307–317.

Vaisey-Genser, M., L.J. Malcolmson, D. Ryland, R. Przybylski, N.A.M. Eskin, and L. Armstrong. 1994. Consumer acceptance of canola oils during temperature-accelerated storage. Food Q. Pref. 5(4):237–243.

Wilks, S.S. 1938. The large sample distribution of the likelihood ratio for testing composite hypotheses. Ann. Math. Stat. 9:60–62.

8 Risk Assessment in Sensory and Consumer Science

8.1. Introduction

Preference mapping is widely used in consumer testing for providing insights regarding what product characteristics are driving differences in overall liking. Internal preference mapping uses consumer data to determine consumer preference patterns, whereas external preference mapping relates consumer data to descriptive sensory information (Lawlor and Delahunty, 2000). Methods known as average external preference mapping are popular because they allow the identification of the relative importance of the sensory attributes toward determining acceptance by consumers. Average external preference mapping has been used for noodles (Tang et al., 2000), rice (Meullenet et al., 2001; Suwansri et al., 2002), salad dressing (Popper et al., 1997), and tortilla chips (Meullenet et al., 2002, 2003). The regression technique used is often partial least squares regression (PLSR). The main advantage of this type of preference mapping is that it is a regression-based technique that allows predictions to be made after the model is developed. In addition, PLSR has the capability of determining the drivers of food acceptance and of comparing the importance of each driver toward overall product liking. One of the major drawbacks of such a technique is the fact that liking scores are averaged across consumers (Popper et al., 1997; Tang et al., 2000). However, liking response patterns by consumers for a set of products are rarely completely homogeneous (Moskowitz and Krieger, 1995). One limitation of PLS regression preference mapping techniques is that the preference mapping model is deterministic; that is, it uses the fixed means of individual input variables (i.e., mean sensory scores to predict consumers' overall acceptance) but does not take into account the variances of the input variables. It is obvious from the second chapter of this text that panel consensus in descriptive analysis is not always achieved and that in some instances, the perception of sensory attributes by panelists can significantly differ without inferring poor performance by the panelists. In such a case, taking the means across panelists and replications may not be a quite truthful representation of the data. In addition, the preference map model can predict the expected value of the output variable but cannot provide the probability of any particular outcome.

Quantitative risk assessment (QRA) is a methodology that can be used to organize and analyze scientific information to estimate the probability and severity of an adverse event. QRA is currently used in food microbiology to identify the health risk in those stages of manufacture, distribution, handling, and consumption of foods associated with foodborne outbreaks (Wei and Fang, 2002). QRA has gained a great deal of popularity with decision makers and analysts in recent years. However, there is no information on the potential use of QRA concepts in the sensory evaluation field.

In this chapter, we consider the application of average preference mapping through PLS regression and its adaptation through the use of the distribution of sensory scores to assess the uncertainty of the predicted acceptance scores.

8.2. Concepts of QRA

Risk assessment is a process to identify a hazard that can have a negative effect and to characterize the risk presented by that hazard. The two basic factors associated with risk assessment are the likelihood of an adverse event and the consequences if it occurs.

Risk is a function of both the probability of an adverse event and the magnitude of the effect of that event. Risk is everywhere in life and business, and especially in new product development. A 1995 survey by the Product Development Management Association indicates that it now takes an average of 6.6 ideas to produce one successful new product and that, on average, 50%–60% of new product development projects failed (Anonymous, 1999). In the case of a new food product launching, there is a potential risk of failure in the market place because of various factors including product sensory quality. For example, if a new food product has undesirable sensory characteristics, this lowers consumer acceptance or increases the rate of rejection of this product. One would expect that product sales would be accordingly affected and that a considerable risk exists in launching this product. There is, therefore, a consumer acceptance of the product (referred as to the standard acceptance score hereafter) below which the product launching results in failure in the market place. Of course, we have to consider that product quality is only one part of the equation predicting product success. However, we will only consider here the product acceptance as the variable determining the risk of product failure.

QRA is widely used in a variety of fields including finance, business, engineering, and food safety. In food microbiology, there are many frameworks for performing a risk assessment, with each framework using a slightly different terminology. In general, microbial quantitative risk assessment involves four distinct steps: hazard identification, exposure analysis, dose–response analysis, and risk characterization (Forsythe, 2000). Hazard identification and risk characterization are applicable to sensory science, but exposure analysis and dose–response analysis may not be applicable. The deterministic model-based stochastic model method is probably most appropriate in sensory science for combining QRA and preference mapping. This is because a deterministic model can be achieved through PLS-based preference mapping and because a stochastic model can be developed by taking into account the random behavior or uncertainty of the input variables. The four steps in the deterministic model-based stochastic model method (@RISK, 2001; Pfeiffer, 1997) are developing a deterministic model, identifying uncertainty for each input variable, developing and simulating a stochastic model, and making a decision.

Developing a deterministic model for the problem of interest is the first step in the application of QRA. In average external preference mapping, a deterministic model can be developed using one of many regression methods such as PLS, principal component regression, proportional odds model, and others (Meullenet et al., 2001, 2002, 2003).

The major disadvantage of a deterministic model is that it only reports averaged expected outcome values and does not provide any information about the likely spread of the outcome values. In most circumstances, the X value controlled in an experiment is likely to be known to higher precision, but this is not always true (Kirkup, 2002), especially in descriptive analysis in which the true means of descriptive attributes (input variables) are uncertain.

Identifying uncertainty for each input variable requires describing the nature of the uncertainty if the input variable is uncertain or stochastic. This is done with probability distributions, which gives both the range of values that the variable could take (minimum to maximum) and the likelihood of occurrence of each value within the range. There are over 30 probability distributions available for describing uncertainty of input variables, including uniform, triangle, normal, lognormal, binomial, beta, and the like. The selection of a distribution for an input variable depends on the problem and data available. Triangle distribution requires only three pieces of information (minimum value, most likely value, and maximum value) and is very useful when you do not have sufficient data to determine the exact distribution. In addition to being certain or uncertain, input variables in the model can be either independent or dependent. An independent variable is totally unaffected by any other variables within the model. A dependent variable, in contrast, is determined in full or in part by one or more other variables in the model. Dependent variables would all be correlated with each other. It is very common in sensory evaluation for sensory attributes to be correlated with each other. It is extremely important to correctly recognize correlations between input variables in QRA, as otherwise, nonsensical results would be generated (@RISK, 2001). Both Pearson's and Spearman's correlations can be used in QRA, depending on their availability from the risk analysis software program. The @RISK program (evaluation version 4.5, Palisade Corporation, New York) uses the Spearman's rank correlation coefficients to describe the correlations among input variables in a stochastic model.

Developing and simulating a stochastic model requires converting the deterministic model into a stochastic model. Once the model and input variables with their distributions are specified, the model can be simulated by methods such as Monte Carlo simulation. Simulation in a sense refers to a method whereby the distribution of possible outcomes is generated by recalculating the model over and over again, each time using a different randomly selected set of input values from the probability distributions. In effect, all valid combinations of the input variable values are tried to simulate all possible outcomes.

Making a decision is the last step. A decision maker would interpret a single-valued result from a traditional analysis as an "expected" value (@RISK, 2001). Most decision-making compares the expected result to some standard or minimum acceptable value. If the result is at least as good as the standard, the researchers then find the result acceptable (@RISK, 2001). However, most would agree that the expected result does not show the effects of uncertainty. One has to somehow manipulate the expected result to make some allowance for risk (@Risk, 2001). For example, one could arbitrarily raise the minimum acceptable result or might weight the chances that the actual result could exceed or fall short of the expected result. At best, the analysis might be extended to include several other results, such as "worst case" and "best case," in addition to the expected value (@RISK, 2001). In QRA, the results are presented in the form of probability distributions that provide a complete picture of all the possible outcomes. As a result, the decision maker no longer just compares desirable outcomes with undesirable outcomes. Instead, one can recognize the range of possible outcomes and that some outcomes are more likely to occur than others. These probability distributions must be interpreted and a decision made based on the interpretation. The same results given to different individuals may be interpreted differently and lead to different courses of action (@RISK, 2001). This is not a weakness of QRA but a direct result of the fact that different individuals have different preferences with regard to possible choices, time, and risk. One individual might feel that the shape of the output distribution shows that the chances of an undesirable outcome far outweigh the chances of a desirable outcome. Another individual who is less risk averse might come to the opposite conclusion.

8.3. A Case Study: Cheese Sticks Appetizers

QRA was applied to the study of cheese sticks appetizers described in Chapter 1. Briefly, the study involved consumer testing of eight products (A–H) at two sites with a total of 150 respondents. In addition, the products were profiled for sensory properties by a trained panel. We consider here the prediction of overall liking and that of the overall rejection rate from either the hedonic response or liking level for appearance, flavor, and texture (internal data from consumers) or the sensory descriptive data (external data from a trained panel of assessors). For the first part of the analysis, which consisted of assessing a deterministic model for predicting either overall acceptance or the rejection rate, the consumer data were averaged across consumers, and the descriptive data were averaged across panelists and replications. The resulting data matrix had eight rows, with each one corresponding to mean values for a given product. Predictive models for overall acceptance and overall rejection rate, defined as the percentage of consumers expressing a dislike for each product for the overall liking question were developed using PLS regression in combination with jackknifing, which allows the selection of significant X variables. The overall rejection rate and overall acceptance were used as the response variables in the models. The overall rejection rate is defined as the percentage of consumers who disliked the product, whereas the overall acceptance is defined as the mean of consumer overall acceptance scores for the product. The PLS predictive models are given by

$$Y = \beta_0 + \Sigma \beta_i x_i + \varepsilon, \qquad (Eq.\ 8.1)$$

where Y is the response variable (overall rejection rate or overall acceptance), x_i is the ith input variable from consumer attribute liking (appearance, flavor, and texture), or sensory attributes ($i = 1 \ldots n$), $\beta_0, \beta_1, \beta_2, \ldots, \beta_n$ are the regression coefficients and ε is the random error. Only significant input variables selected by the jackknife optimization method were included in the PLS predictive models. Full-cross validation was used to assess the robustness of the predictive models. The goodness of fit of a predictive model was assessed using calibration coefficient of determination R^2 and root mean square of error of prediction (RMSEP). The optimized PLS predictive models were obtained using the Unscrambler multivariate software package (version 9.0, CAMO ASA, Norway). Because it is assumed that there are no errors in the values of input variables (X) in the least square method (Kirkup, 2002), the PLS models are deterministic.

The predictive models were then inputted into @Risk, a QRA modeling software. In addition, data from descriptive flavor and texture attributes for each product were fitted with normal distributions and the distribution parameters used as the input distributions. For descriptive appearance attributes, triangle distributions were obtained from the panel consensus data. The correlation structure among the predictive variables was also inputted in @Risk before the simulation was launched. The @RISK software package was used with a simulation iteration number of 10,000 and a Latin square hypercube sampling for the simulation. Regression Tornado Graphs were used for sensitivity analysis.

8.3.1. Prediction of Overall Rejection Rate from Attribute Acceptance Data

The rejection rates for overall impression, appearance, flavor, and texture acceptance are provided in Table 8.1. The table shows that product B had the lowest overall rejection rate, and product C had the highest overall rejection rate. The consumer overall rejection

Table 8.1. Percentages of consumers who disliked, neither disliked nor liked, and liked the fried cheese stick appetizer products.

Product	Overall Dislike	Overall Neither	Overall Like	Appearance Dislike	Appearance Neither	Appearance Like	Flavor Dislike	Flavor Neither	Flavor Like	Texture Dislike	Texture Neither	Texture Like
A	20.50	6.83	72.67	14.29	4.97	80.75	22.98	3.73	73.29	25.47	9.94	64.6
B	14.11	3.68	82.21	7.98	4.29	87.73	23.31	5.52	71.17	25.77	4.91	69.33
C	57.41	6.17	36.42	59.26	9.88	30.86	48.77	7.41	43.83	65.43	10.49	24.07
D	15.53	5.59	78.88	8.7	6.21	85.09	23.6	6.83	69.57	17.39	8.70	73.91
E	32.1	6.79	61.11	28.4	4.32	67.28	41.36	6.79	51.85	30.25	8.02	61.73
F	45.34	6.83	47.83	44.1	9.32	46.58	49.69	9.32	40.99	44.72	9.32	45.96
G	32.92	7.45	59.63	49.69	6.83	43.48	31.68	9.32	59.01	29.81	9.94	60.25
H	20.86	4.29	74.85	13.5	3.68	82.82	17.18	3.07	79.75	26.99	7.36	65.64

rate (Y) was modeled using the rejection rates of consumer appearance (A), flavor (F), and texture (T) as predictors. The fitted PLS predictive model is given here:

$$Y = -0.03150 + 0.26775A + 0.41452F + 0.36226T. \qquad \text{(Eq. 8.2)}$$

The R^2 and RMSEP for this predictive PLS model were 0.98 and 0.029, respectively, indicating that the PLS predictive model fitted very well. The standardized regression coefficients of a PLS predictive model allowed the assessment of the relative importance of each input variable to the prediction of the response variable (Meullenet et al., 2001). In this case, the standardized regression coefficients for the rejection rates of appearance, flavor, and texture were 0.355, 0.344, and 0.358, respectively, which indicated that these three variables were nearly equally important to the overall rejection rate, with the texture rejection rate being slightly more important.

The predicted overall rejection rates for the eight products studied are listed in Table 8.2. If the standard or maximum acceptable overall rejection rate was 30%, then products A, B, D, and H were acceptable products because their estimated overall rejection rates were less than 30%, and the remaining products were not acceptable. This inference from the PLS model was straightforward, which is an advantage of the deterministic model. However, this deterministic model cannot estimate the probability of product A having an overall rejection rate of 30% or more. In addition, the result from the deterministic PLS model is based on the assumption that the true means of the rejection rates for appearance, flavor, and texture are certain and known. In reality, those true means are uncertain and unknown. Given the information available, the ranges of the true means for the eight products could be best described by the triangle distributions listed in Table 8.3. To convert the deterministic model to a stochastic model, it is usually assumed that the deterministic model is considered to adequately represent the underlying problem, and then single value estimates of the input variables are replaced with probability distributions. For example, the following stochastic model for product A could be derived from the PLS deterministic model by replacing the input variables with their triangle distributions if one ignores the correlations among the input variables:

$$\begin{aligned}Y = &-0.03150 + 0.26775 Triangle(0.0932, 0.1429, 0.1926) + \\ & 0.41452 Triangle(0.1925, 0.2298, 0.2671) + \\ & 0.36226 Triangle(0.1552, 0.2547, 0.3541).\end{aligned} \qquad \text{(Eq. 8.3)}$$

The results from the Monte Carlo simulation on the stochastic models showed that the mean rejection rate value (19.43%) for product A has a 15.2% chance of occurring and that the range of the overall rejection rate is estimated to be between 13.12% and 25.78% (Table 8.2). It is, however, certain that the overall rejection rate is less than 26% if the underlying deterministic model is considered correct. The minimum, mean, and maximum values for the other products, as well as the observed values, are listed in Table 8.2. Results clearly show that the model-estimated and observed overall rejection rates were similar, implying that the model appeared to be appropriate.

The distributions of the overall rejection rates for the eight products are presented in Figure 8.1a. The figure reflects how likely it is for a product to exhibit a certain rejection rate, taking into account the variation of the consumer data obtained. The y axis represents how frequently values along the x axis were estimated to occur during the simulation of the model. Different products had different peaks in their output distributions. The peak of an output distribution represents the probability of having the averaged overall

Table 8.2. Estimated minimum, mean, and maximum values for the overall rejection rate for eight products (A–H).

Product	Observed Overall Rejection Rate (%)*			Estimated Overall Rejection Rate (%) from Internal Data**			Estimated Overall Rejection Rate (%) from External Data***		
	Minimum	Mean	Maximum	Minimum	Mean	Maximum	Minimum	Mean	Maximum
A	13.67	20.50	27.33	13.12	19.43	25.78	0	21.14	59.17
B	10.43	14.11	17.79	12.96	17.98	23.08	0	11.71	48.88
C	51.24	57.41	63.58	47.49	56.64	66.08	0.03	43.18	78.25
D	9.94	15.53	21.12	7.85	15.26	22.68	0	17.11	48.72
E	25.31	32.10	38.89	25.87	32.56	39.30	0.08	44.33	71.07
F	38.51	45.34	52.17	35.90	45.46	54.80	0.13	38.40	66.74
G	25.47	32.92	40.37	25.03	34.09	43.22	0	33.89	69.53
H	16.57	20.86	25.15	12.54	17.36	22.11	0	18.77	54.75

*The observed rejection rate was calculated from the mean of the overall rejection rate minus or plus the portion of undecided consumers.
**The overall rejection rate was estimated with @Risk, using appearance, flavor, and texture rejection rates as the X-variables.
***The overall rejection rate was estimated with @Risk, using significant sensory attribute scores as the X-variables.

Table 8.3. Triangle distributions of the rejection rates for eight fried cheese sticks appetizer products.

Product	Input Variable	Distribution
A		
	Appearance	Triangle(0.0932, 0.1429, 0.1926)
	Flavor	Triangle(0.1925, 0.2298, 0.2671)
	Texture	Triangle(0.1553, 0.2547, 0.3541)
B		
	Appearance	Triangle(0.0369, 0.0798, 0.1227)
	Flavor	Triangle(0.1779, 0.2331, 0.2883)
	Texture	Triangle(0.2086, 0.2577, 0.3068)
C		
	Appearance	Triangle(0.4938, 0.5926, 0.6914)
	Flavor	Triangle(0.4136, 0.4877, 0.5618)
	Texture	Triangle(0.5494, 0.6543, 0.7592)
D		
	Appearance	Triangle(0.0249, 0.0870, 0.1491)
	Flavor	Triangle(0.1677, 0.2360, 0.3043)
	Texture	Triangle(0.0869, 0.1739, 0.2609)
E		
	Appearance	Triangle(0.2408, 0.2840, 0.3272)
	Flavor	Triangle(0.3457, 0.4136, 0.4815)
	Texture	Triangle(0.2223, 0.3025, 0.3827)
F		
	Appearance	Triangle(0.3478, 0.4410, 0.5342)
	Flavor	Triangle(0.4037, 0.4969, 0.5901)
	Texture	Triangle(0.3540, 0.4472, 0.5404)
G		
	Appearance	Triangle(0.4286, 0.4969, 0.5652)
	Flavor	Triangle(0.2236, 0.3168, 0.4100)
	Texture	Triangle(0.1987, 0.2981, 0.3975)
H		
	Appearance	Triangle(0.0982, 0.1350, 0.1718)
	Flavor	Triangle(0.1411, 0.1718, 0.2025)
	Texture	Triangle(0.1963, 0.2699, 0.3435)

rejection rate for the product. The probabilities of having the averaged overall rejection rates for products A–H to occur were 15.2%, 18.7%, 10.7%, 12.9%, 15.1%, 10.5%, 11.4%, and 18.7%, respectively. In traditional deterministic modeling, however, it is assumed that these probabilities are 100%, which is not correct from a risk-assessment point of view. The figure also shows different shapes of the output distributions for the products. Product B had a narrow and sharp distribution shape, whereas product F exhibited a flat and wide distribution. The flatter the distribution is, the greater the uncertainty of the results. Therefore, results for product F were more uncertain than were those for product B, implying that people can be more certain of the decision they may make on product B than of the decision made on product F. Products A, B, C, and D had similar mean overall rejection rates. However, there is more uncertainty about the rejection rates for products A and D than for products B and C.

The cumulative probability plots are presented in Figure 8.1b. These plots allow people to very quickly obtain the probability of the outcome variables being greater than a particular value. They convey the same information as a table of percentiles for a particular distribution. For example, if an acceptable rejection were 30% (i.e., the standard overall rejection rate), product A was acceptable because the probability of the overall rejection

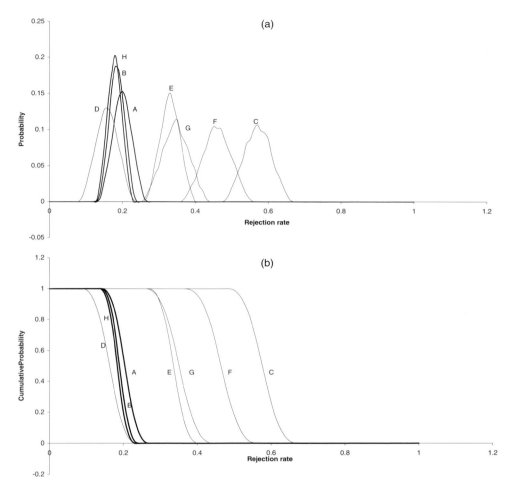

Figure 8.1. Plots of probability and cumulative probability for the risk of the overall rejection rates for products A–H: (a) probability density curves and (b) cumulative probability density curves.

rate being greater than 30% was zero. If the standard overall rejection rate was 20%, then the probability of the overall rejection rate being greater than 20% was 56.43%. Whether this product is acceptable would be dependent on the decision maker's preference for a risk level. This is an advantage of QRA over the traditional approach.

The use of uncorrelated or correlated input variables has a significant effect on the risk distribution of the output variable. In this case, the rejection rates for appearance, flavor, and texture were correlated with each other ($p < 0.05$; Table 8.4). A comparison of the simulated rejection rate distribution when either using or not using the correlation structure is given in Figure 8.2 for product C. It is apparent that the output distribution was much wider and flatter when including the correlation structure in the model than when excluding it. This implies that the correlation structure should be used to simulate the distribution of hedonic variables.

Sensitivity analysis was performed to determine which input variables had more influence on determining the risk of the overall rejection rate. The results from sensitivity analysis are presented in Figure 8.3. The figure shows that texture was the most important

Table 8.4. Spearman rank correlation coefficients of the rejection rates for appearance, flavor, and texture acceptance (significantly different from zero at $p < 0.05$).

	Appearance Rejection Rate	Flavor Rejection Rate	Texture Rejection Rate
Appearance rejection rate	1	0.6905	0.8095
Flavor rejection rate		1	0.7619
Texture rejection rate			1

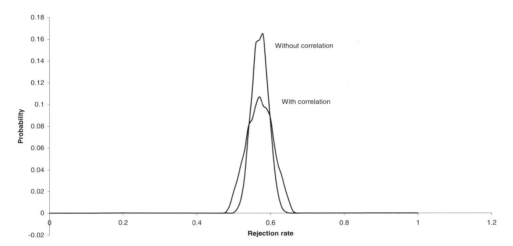

Figure 8.2. Comparison of probability of overall rejection rate for product C both with and without the correlation structure included in the model.

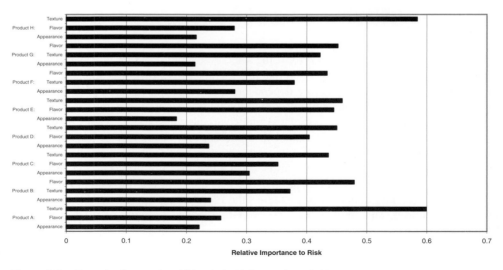

Figure 8.3. Tornado diagram (sensitivity analysis) for products A–H.

factor for the overall rejection rate for five of the eight products, whereas flavor was most important for three of the eight products. Therefore, one could conclude that reducing the texture or flavor rejection rates will most reduce the rate of risk of overall rejection for the eight products tested. The figure also shows that different products had different sensitivities in the risk distribution of the overall rejection rate. For example, product A had a sensitivity value of 0.599 for texture, whereas product B had a sensitivity value of 0.372. A sensitivity value of 0.599 means that an increase of one standard deviation in texture rejection rate increases the overall rejection rate by 0.599 standard deviations.

8.3.2. Prediction of Overall Rejection Rate from Descriptive Sensory Attributes

The same procedure was applied to the prediction of the overall rejection rate from sensory attributes measured by the trained panel. The consumer overall rejection rate (Y) was modeled using 44 descriptive attributes as predictors in a PLS regression model. The optimized PLS predictive model is given here:

$$Y = 0.7708 + 0.114084C - 0.11216O - 0.06603H - 0.02162S, \quad \text{(Eq. 8.4)}$$

where C is the aftertaste cardboard, O is the onion/garlic flavor, H is the hardness of cheese, and S is the size uniformity. The R^2 and RMSEP for this predictive PLS model were 0.99 and 0.031, respectively. The PLS standardized regression coefficients for the aftertaste cardboard, onion/garlic, hardness of cheese, and uniformity of size were 0.317, −0.322, −0.346, and −0.489, respectively, indicating that the uniformity of size had more influence on the overall rejection rate than did the other attributes. Increased cardboard aftertaste tended to increase the overall rejection rate, whereas increased onion/garlic flavor, cheese firmness, and size uniformity all reduced rejection rates.

The PLS model was converted into a stochastic model by specifying a distribution for each of the four input variables (Table 8.5). The input distributions in Table 8.5 were determined from the descriptive data. Because there were no significant correlations among the four input variables ($p < 0.05$), no correlation structure was included in the stochastic model. The Monte Carlo simulation results are presented in Figure 8.4. The output distributions (Figure 8.4) were not smooth because @RISK only allows a maximum iteration of 10,000 for Latin square hypercube sampling simulations. The minimum, mean, and maximum overall rejection rates estimated by the stochastic model are listed in Table 8.2. The model, on the basis of external data, estimated means for the overall rejection rates of the various products similar to those predicted from the internal data model. However, the range of predicted overall rejection rates (i.e., minimum and maximum)

Table 8.5. Distributions of input variables in the partial least squares models based on external data.

Product	Aftertaste Cardboard	Onion/Garlic	Hardness of Cheese	Uniformity of Size
A	Normal(2.82, 0.94)	Normal(3.64, 0.68)	Normal(3.63, 1.13)	Triangle(7, 9, 11)
B	Normal(3.00, 0.87)	Normal(3.37, 0.78)	Normal(3.20, 1.39)	Triangle(10.5, 12.5, 14.5)
C	Normal(2.76, 0.84)	Normal(3.74, 1.20)	Normal(2.17, 1.33)	Triangle(8, 10, 12)
D	Normal(3.00, 0.87)	Normal(3.66, 0.73)	Normal(2.73, 1.24)	Triangle(11, 13.5, 15)
E	Normal(3.60, 0.58)	Normal(3.62, 0.47)	Normal(2.60, 1.39)	Triangle(7, 7, 9)
F	Normal(2.75, 0.90)	Normal(3.76, 0.71)	Normal(2.14, 0.80)	Triangle(6, 8.5, 11)
G	Normal(2.68, 1.03)	Normal(3.52, 0.61)	Normal(3.44, 1.24)	Triangle(6, 8, 10)
H	Normal(3.20, 0.80)	Normal(4.15, 1.01)	Normal(4.49, 1.53)	Triangle(7, 9.5, 12)

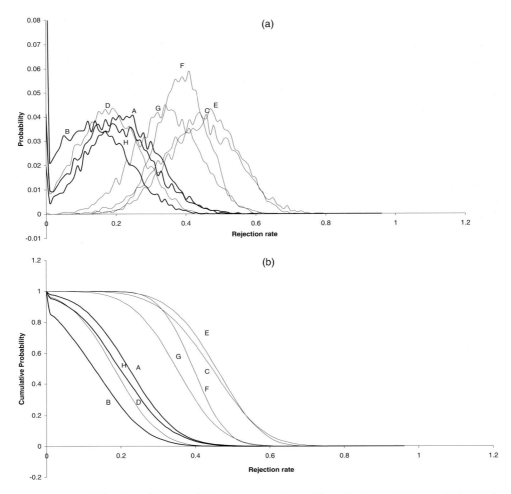

Figure 8.4. Distributions of the overall rejection rates estimated from the external data model for products A–H: (a) probability density curves and (b) cumulative probability density curves.

estimated using the external data was much wider than that using the internal data model (Table 8.2). This was because the descriptive attributes score distributions were broader than the distributions of the consumer rejection rates for appearance, flavor, and texture. This implies that use of descriptive attributes to predict consumer overall rejection rate may result in more widely spread output distributions and more difficult decision making. However descriptive analysis provides greater details about the specific sensory modalities driving product liking or rejection.

8.3.3. Prediction of Overall Liking from Descriptive Sensory Attributes

As discussed previously, the overall rejection rate was used for the application of QRA because QRA is mostly used to identify the likelihood of adverse events. In practice, however, it is rarely, if at all, used in the preference mapping framework. As a result, it may be more appropriate for sensory scientists to apply this type of simulation to the prediction of overall liking. The optimized PLS predictive model for overall acceptance is given here:

$$Y = 3.24545 - 0.61822C + 0.46622O + 0.40209H + 0.12637S. \quad \text{(Eq. 8.5)}$$

The R^2 and RMSEP for this predictive PLS model were 0.998 and 0.119, respectively, indicating that the PLS predictive model fitted well. The PLS standardized regression coefficients for the aftertaste cardboard, onion/garlic, hardness of cheese, and uniformity of size were −0.314, 0.244, 0.384, and 0.521, respectively. As expected, the results were similar to those from the external preference map for the overall rejection rate. The estimated means of the overall acceptance for products A–H by the PLS model were 6.19, 6.74, 5.09, 6.48, 4.98, 5.32, 5.6, and 6.52, respectively. According to the deterministic PLS model, a standard overall acceptance score of 6.5 (i.e., the minimum acceptable mean) would result in only products B and H being acceptable.

The Monte Carlo simulation results are presented in Figure 8.5. The probability density curves (Figure 8.5a) were not smooth because of the limitations of the @RISK program. The odds of products A–H having an overall acceptance means greater than 6.5 (the standard overall acceptance score) were 0.2945, 0.6461, 0.0127, 0.4805, 0.0052, 0.0019,

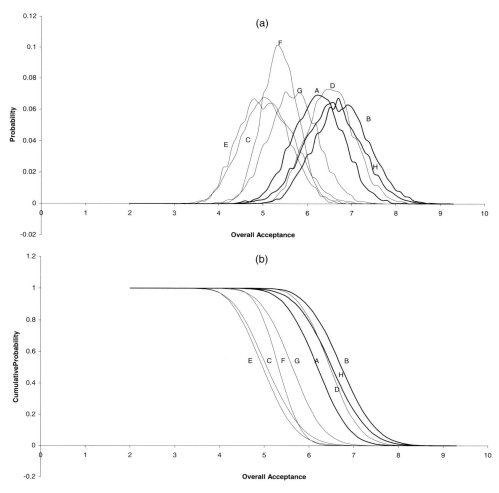

Figure 8.5. Distributions of the overall acceptance estimated from the external data model for products A–H: (a) probability density curves and (b) cumulative probability density curves.

0.062, and 0.5124, respectively (Figure 8.5b). The identification of acceptable products would then depend on one's interpretation of the results. If the decision maker used a minimum acceptable risk of 70% as the decision rule, then no products would be deemed acceptable. If the minimum acceptable risk was 60%, then only product B would be acceptable. If the standard overall acceptance scores were changed to 6.0, then the probabilities for the overall acceptance means being greater than 6.0 were 0.632, 0.882, 0.072, 0.812, 0.0340, 0.043, 0.253, and 0.791, respectively. Considering a minimum acceptable risk to be 70% results in products B, D, and H being acceptable. This implies that the final decision from QRA depends on two criteria: the minimum acceptable value for the outcome and the maximum acceptable risk for the occurrence of the outcome.

8.4. Conclusions

QRA using Monte Carlo simulation offers a powerful method for assimilating the various uncertainties of a problem and producing an accurate QRA model. The foundation block of this simulation technique is the underlying probability distributions of input variables. QRA characterizes the nature and likelihood of an adverse event, helps to define the uncertainties, and provides some level of comfort with the inferences that are being made. However, there are some weaknesses to the risk assessment process as well. Risk assessment has its roots in toxicology and carcinogenicity studies, and its application to other disciplines poses significant challenges. For food sensory scientists, there is a difficulty in defining the adverse event for QRA. In addition, QRA is subject to two types of uncertainties: those related to data and those associated with any assumptions that are required when directly applicable data are not available. QRA can help document and clarify the components of risk, given the information that is available at that time, leading to a more efficient and effective utilization of resources and better decisions. QRA, as are other techniques that will yield results indicating uncertainty or risk, could be an emerging tool for sensory scientists to estimate the risk of consumer rejection of new products introduced to the market place. Further research needs to be focused on the effects of the uncertainty of the model's parameters on the output distributions.

References

Anonymous. 1999. Corporate new product project portfolio design. Available from http://www.sensors-research.com/rsmstart.htm. Accessed on September 30, 2002.
Risk. 2001. Guide to using @RISK: risk analysis and simulation add-in for Microsoft Excel (version 4.0). New York: Palisade Corporation.
Forsythe, S.J. 2000. The microbiology of safe food. Malden: Blackwell Science.
Kirkup, L. 2002. Data analysis with Excel: An introduction for physical scientists. Cambridge: Cambridge University Press.
Lawlor, J.B., C.M. Delahunty. 2000. The sensory profile and consumer preference for ten speciality cheeses. Int. J. Dairy Technol. 53:28–36.
Meullenet, J.-F., V.K. Griffin, K. Carson, G. Davis, S. Davis, J. Gross, J.A. Hankins, E. Sailer, C. Sitakalin, S. Suwansri, and A.L. Vasquez Caicedo. 2001. Rice external preference mapping for Asian consumers living in the United States. J. Sensory Stud. 16:73–94.
Meullenet J.-F., R. Xiong, M.A. Monsoor, T. Bellman-Homer, P. Dias, S. Zivanovic, H. Fromm, and Z. Liu. 2002. Preference mapping of commercial toasted white corn tortilla chips. J. Food Sci. 67:1950–1957.
Meullenet, J.-F., R. Xiong, J.A. Hankins, P. Dias, P. Zivanovic, M.A. Monsoor, T. Bellman-Homer, Z. Liu, and H. Fromm. 2003. Preference modeling of commercial toasted white corn tortilla chips using proportional odds model. Food Q. Pref. 14(2003):603–614.

Moskowitz, H.R. and B. Krieger. 1995. The contribution of sensory liking to overall liking: An analysis of six food categories. Food Q. Pref. 6:83–90.

Pfeiffer, D.U. 1997. Quantitative risk assessment. In R. Rupanner (ed.). Risk analysis and animal health—A course manual. International Training Course, Dubendorf, Switzerland, July 13–18, pp. 701–721.

Popper, R., H. Heymann, and F. Rossi. 1997. Three multivariate approaches to relating consumer to descriptive data. In A.M. Munoz (ed.). ASTM manual on consumer data relationships. Philadelphia: ASTM, pp. 39–61.

Suwansri, S., J.-F. Meullenet, J.A. Hankins, and V.K. Griffin. 2002. Preference mapping of jasmine rice for U.S. Asian consumers. J. Food Sci. In Press.

Tang, C., H. Heymann, and F. Hsieh. 2000. Alternatives to data averaging of consumer preference data. Food Q. Pref. 11:99–104.

Wei Q.K., and T.J. Fang. 2002. Quantitative risk assessment for Listeria monocytogenes in contaminated cooked meat. 2002 IFT Annual Meeting Technical Program (abstracts), Anaheim, California, June 15–19.

9 Application of MARS to Preference Mapping

9.1. Introduction

When mapping the relationship between hedonic ratings and analytical data (such as descriptive or instrumental data), one of the main problems is that the relationship between these variables is expected to be complex or nonlinear and to involve interaction effects. Because we usually know little about the nature of these relationships, preference mapping is usually a rather complex and exploratory process.

When the relationship between a hedonic response and predictors is complex or nonlinear, the use of typical linear PLS (partial least squares) regression models may result in misleading interpretations and conclusions. An alternative to modeling such complex relationship is MARS (multivariate adaptive regression splines; Salford Systems, 2001). MARS, developed by Stanford physicist and statistician Jerome Friedman (1991), is a multivariate nonparametric regression method that can be applied to model the relationships between intensity sensory ratings or diagnostic JAR and hedonic data. MARS uses smoothing splines to determine the relationship between a vector of predictors and a response variable (Sephton, 2001). MARS is based on partitioning the predictor variable space into separate regions, each of which gets its own regression line (Steinberg, 2001). MARS is a data-driven procedure compared to the more frequently used model-driven procedures. The MARS model is a regression model that uses basis functions as predictors in place of the original predictive variables, similar to the use of principal components in PLS and PC (principal component) regressions in place of the original variables.

The "optimal" MARS model, a linear combination of basis functions, is selected in a two-phase process (forward and backward phases; Salford Systems, 2001). In the forward phase, a model is grown by adding basis functions until an overly large model is found. In the backward phase, basis functions are deleted in the order of least contribution to the model until an optimal balance of bias and variance is found. By allowing for any arbitrary shape of the response function as well as for interactions, and by using the two-phase model selection method, MARS is capable of reliably tracking very complex data structures that often hide in high-dimensional data.

MARS has been proven effective at a variety of learning problems and is competitive with neural networks and non-parametric regressions (Dwinnell, 2000). Xiong and Meullenet (2004) have applied MARS to consumer preference mapping, using consumer test data for fried mozzarella cheese sticks. Their results show that MARS has potential for modeling the complex relationship between liking and analytical data in the preference mapping framework.

9.2. MARS Basics

To apply MARS to sensory science problems, one needs to first understand some basic terminology used in MARS. These include knots, basis functions, piecewise linear regression splines, and GCV (generalized cross validation). First, let us take a simple example comparing MARS and linear regression models before we introduce the MARS terminology. Suppose that a sensory analyst wants to examine the relationship between overall liking and purchase intent for a product. He has used 157 consumers in his consumer study and obtained the scatter plot of overall liking (on the 9-point hedonic scale, with 1 = dislike extremely and 9 = like extremely) versus purchase intent (on a 5-point scale, with 1 = definitely would not buy and 5 = definitely would buy) given in Figure 9.1.

Figure 9.1a clearly shows that there are some outliers in the original data set. After examining the original data, the analyst found two outliers; the data without outliers are presented again in Figure 9.1b. The relationship between overall liking and purchase intent is relatively simple in this case; that is, purchase intent obviously increases as overall

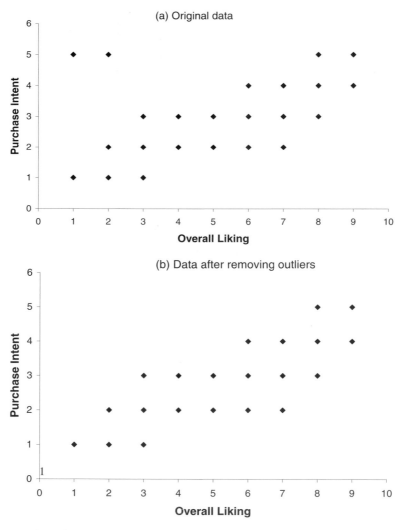

Figure 9.1. Relationship between overall liking and purchase intent (N = 157).

liking increases. There are several statistical procedures that can be used to determine the relationship. One of the most common procedures for curve fitting is simple linear regression. Simple linear regression assumes that the relationship between the response and explanatory variables is a straight line. Figure 9.2a displays the graph of a fitted straight line of purchase intent over overall liking. The simple linear model has the advantage of easy interpretation (i.e., purchase intent increases by 0.4772 as a 1-unit increase of overall liking in this example) but is not accurate if the true relationship is in fact nonlinear. Another statistical procedure that is frequently used in classical linear regression to approximate nonlinear curves is polynomial regression. Figure 9.2b displays a graph of a fitted quadratic polynomial regression model. A quadratic model usually approximates the "true" relationship better than a simple linear model. In this example, an F test shows that

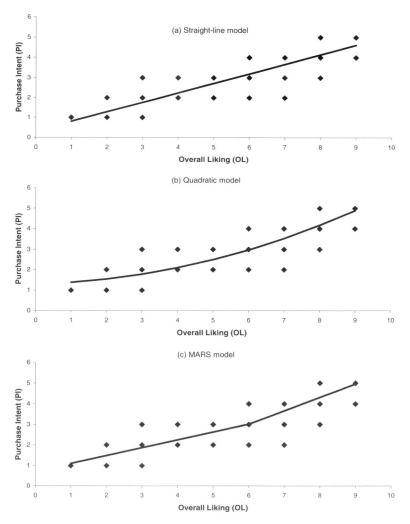

Figure 9.2. Various models fitted to the relationship between overall liking and purchase intent. Linear model: PI = 0.3255 + 0.4772 × OL; quadratic model: PI = 1.3015 + 0.0407 × OL + 0.0397 × OL; MARS model: PI = 3.014 + 0.648 × BF1−0.381 × BF2, where BF1 = Max(0, OL-6) and BF2 = Max(0, 6-OL).

the quadratic model fits statistically significantly better that the straight-line model. The quadratic model indicates that purchase intent increases at a slower rate over the overall liking region of 1–6 than over the region of 6–9. Because the slopes of the fitted quadratic curve for purchase intent change over the region of 1–9, interpretation of regression coefficients in the quadratic model is more difficult. The use of polynomial regression to approximate any functions is familiar to readers from its use in Taylor series expansions. However, the Taylor series expansion is used to approximate a function near a point rather than over a wide range. When evaluating a function over a range, the maximums and inflection points of the polynomial may not exactly match the curves of the function being approximated (Francis, 2003).

The third statistical procedure applied in this example is the MARS approach, a nonparametric regression model. The MARS approach to fitting nonlinear functions has similarities to polynomial regression. In its simplest form, MARS fits piecewise linear regressions to the data. That is, MARS breaks the data into regions and fits different straight lines (local linear regression models or piecewise linear regression models) for the different regions. MARS requires the function fit to be continuous, and thus there are no jump points between contiguous regions (Francis, 2003). Figure 9.2c displays the fitted MARS model, which breaks the whole data space into two ranges at the overall liking score of 6 and fits two straight lines with different slopes for the two ranges. The slope (0.381) of the first straight line over the overall liking region of 1–6 is less than that (0.648) of the second straight line over the region of 6–9, which is similar to the polynomial regression model. The interpretation of the two slopes for the MARS model is much easier than for the polynomial regression model. According to the MARS model, the overall liking score of 6 corresponds to the purchase intent of "either probably buy or probably not buy." As overall liking score increases from 1 to 6, purchase intent increases at a slower rate of 0.381 per unit hedonic score, which seems reasonable because consumers have lower purchase intent for products they dislike. As overall liking increases from 6 to 9, purchase intent increases at a higher rate of 0.648 per unit hedonic score, indicating that consumers have higher purchase intent for the products they like. One can see the score of 6 as a threshold for purchase. Below a score of 6, purchase intent is low and does not increase much. However, above a score of 6, purchase intent grows more rapidly with increasing overall liking. Compared to the simple linear model (straight line), the MARS model seems to provide more information about the relationship. In addition, MARS is more easily interpreted than the quadratic model.

This example illustrates the fact that traditional techniques requiring specific assumptions about the relationship between dependent and independent variables may lack the flexibility to model nonlinear relationships. It should be noted, however, that when the true relationship between the dependent and independent variables is linear, classical statistical methods are likely to outperform complex models such as neural networks.

Linear and quadratic models are global because they apply to the whole data region, whereas MARS models are local because they apply to different regions. To partition the data range into several local regions, one needs to understand the concept of knots.

9.2.1. Knots

A key concept underlying the concept of splines is the knot, which marks the end of one region of data and the beginning of another. Thus, the knot is where the behavior of the function changes. In a classical spline, the knots are predetermined and evenly spaced, whereas in MARS the knots are determined by a search procedure, depending on the data. Only the minimum number of knots necessary to describe the relationship between two variables is included in a MARS model. If a straight line is a good fit, there will be no interior knots. In MARS, however, there is always at least one "pseudo" knot that corresponds to the smallest observed value of the predictor (Salford Systems, 2001). Knots are found via an exhaustive, brute force search procedure using fast-update algorithms. For example, Figure 9.3a shows the scatter plot of observed data between hedonic liking response Y and intensity variable X (artificial data). The relationship between Y and X indicates an optimal intensity level at which the liking score is maximal. This is a typical optimization problem faced by sensory professionals, which is traditionally solved using a polynomial regression model (i.e., a quadratic model) to find the optimal solution. MARS, however, can be used as an alternative method to solve this type of sensory optimization problem. On the basis of the data, MARS found three knots (X_1, X_2, and X_3) that partition the whole data space into four local regions (Figure 9.3b). Within each local region, MARS fits a straight line, and the slopes of the two straight lines at each knot are changed. Four local straight lines together describe the nonlinear relationship between Y and X.

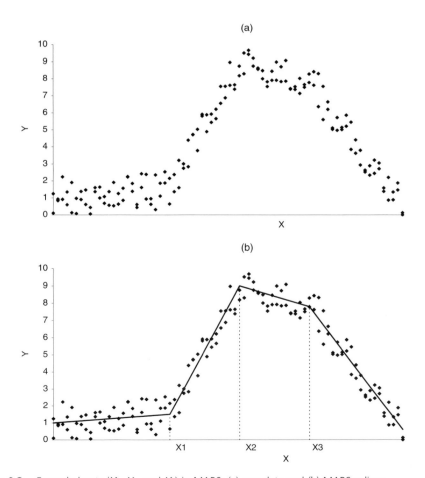

Figure 9.3. Example knots (X_1, X_2, and X_3) in MARS: (a) raw data and (b) MARS splines.

9.2.2. Basis Functions

"Basis function" is a new term coined for MARS. Basis functions are a set of functions used to represent the information contained in one or more variables. In MARS, basis functions are the machinery used for generalizing the search for knots. The basis functions that look like hockey sticks in Figure 9.4 are the core building blocks of the MARS model and are often applied to a single variable multiple times. In actuality, the basis function maps a variable X to a new variable BF (called basis function), using one of the following two step functions:

$$BF = \text{Max}(0, X - C) = \begin{cases} 0 & \text{If } X \leq C \\ X - C & \text{If } X > C \end{cases} \quad \text{(Eq. 9.1)}$$

or

$$BF = \text{Max}(0, C - X) = \begin{cases} C - X & \text{If } X \leq C \\ 0 & \text{If } X > C' \end{cases} \quad \text{(Eq. 9.2)}$$

where $BF = \text{Max}(0, X-C)$ is the standard basis function (see Figure 9.4a), $BF = \text{Max}(0, C-X)$ is the image basis function (see Figure 9.4b), X is an independent variable, and C is a constant (called threshold value) that will be estimated in MARS.

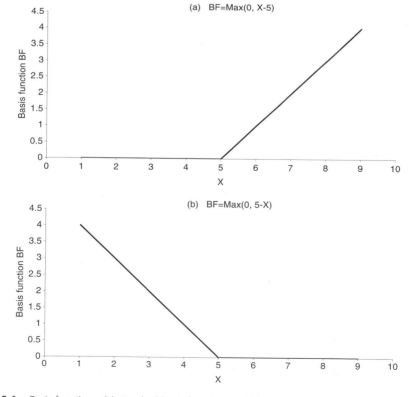

Figure 9.4. Basis functions: (a) standard basis function and (b) image basis function.

Alternative notations in the literature for the basis functions are $(X\text{-knot})_+$ and $(\text{knot-}X)_+$, which have the exact same meaning as Max(0, X-knot) and Max(0, knot-X), respectively. The function max(x_1, x_2) is interpreted as the maximum value of the two elements, x_1 and x_2. The definition of basis functions indicates that locating knots is equivalent to determining the threshold value C in the basis function. The threshold value C is determined by relating the response variable (Y) to basis functions (BF), which is discussed in the following section.

9.2.3. Piecewise Linear Regression Splines

Instead of fitting a single straight line (global model) to the data, MARS fits a series of local straight lines to the subregions of the data (Figure 9.3b). The collection of local straight lines is called piecewise linear regression splines. The piecewise linear regression splines, also called the MARS model, can be expressed as

$$Y = \beta_0 + \beta_1 BF_1 + \beta_2 BF_2 + \ldots + \beta_p BF_p + e, \quad \text{(Eq. 9.3)}$$

where Y is the response variable, BF_i ($i = 1, 2, \ldots, p$, where p is the number of basis functions) is the ith basis function, β_i ($i = 0, 1, 2, \ldots, p$) is regression coefficient for BF_i, and e is the error term. Much like principal components, basis functions essentially reexpress the relationship of the predictor variables with the response variable. The relationship between the response and basis functions shows that any basis function defines a knot where a regression can change slope. Running a regression on basis functions is equivalent to specifying a piecewise linear regression. Thus, the problem of locating knots is now translated into the problem of defining basis functions.

MARS always defines basis functions in pairs (standard and image basis functions) by searching in a stepwise manner. It starts with just a constant in the model and then begins the search for a (X-knot) combination that improves the model the most in terms of the GCV measure. This search is then repeated, with MARS searching for the best variable to add, given the basis functions already in the model. The search process theoretically continues until every possible basis function has been added to the model. This is the forward search phase, in which the MARS deliberately constructs an overly large model by adding basis functions. As more basis functions are added, the model becomes more flexible and more complex in estimating the "true" function, and the process continues until a user-specified maximum number of basis functions is reached. Once forward searching is done, backward searching is immediately performed to prune basis functions that have the least contribution to the model, until an optimal model is found. The rationale for having two search phases is that the good model cannot be built from a forward stepwise plus stopping rule; rather, the model must be generously overfitted and then have the unneeded basis functions removed (Salford Systems, 2001). In practice, however, the model needs to be limited because of the intensity of the search that may be required with large data sets. For example, with 400 variables and 10,000 records, there are 4 million possible knots to examine just for the main effects. Even if most variables have a limited number of distinct values, the total number of possible knots will be very large. The user specifies an upper limit for the number of knots to be generated in the forward stage. The limit should be large enough to ensure that the true relationship can be captured. A good rule of thumb for determining the minimum number is three to four times the number of basis functions in the optimal model (Salford Systems, 2001). This limit is usually set by trial and error.

9.2.4. GCV

This section discusses the way model quality in MARS is assessed. One popular approach used in data mining applications is to partition the entire data set into training and testing data sets. Typically, one-half to one-third of the data is held out for testing. However, when the data set is small, it is difficult to partition it into training and testing data sets and still maintain data sets that are representative. In addition, as the testing is performed on a relatively small data set, the goodness-of-fit results may be sensitive to random variation in the subsets selected for training and testing. An alternative procedure that allows more of the data to be used for fitting and testing is cross-validation (CV), which is a popular method of measuring how well the model will predict future observations. CV involves iteratively holding out part of the data, fitting the model to the remainder of the data, and testing the goodness-of-fit of the model on the held-out portion. For example, the data may be divided into four groups. Three of the groups are used to fit the model, and one is used for testing. The process is repeated four times, and the goodness-of-fit statistics for the four test samples are averaged. Extending the CV concept, GCV uses each data point as the testing set and then averages a weighted prediction error over the entire data set (Golub et al., 1979). Because CV requires a complete reconstruction of the model for each observation, the CV score (i.e., the average squared difference between each observed data point and the model recalculated with that observation removed) is usually too expensive to compute. GCV was introduced by Grace Wahba (Craven & Wahba 1979) as an approximation to CV and can be expressed as

$$GCV = \frac{1}{N} \frac{\sum_{i=1}^{N}(y_i - \hat{f}(x_i))^2}{\left(1 - \frac{df}{N}\right)^2}, \quad \text{(Eq. 9.4)}$$

where N is the number of observations on which the model is built, y_i is the ith observation of the dependent variable y, x_i is the ith observation of the independent variable x, $\hat{f}(x)$ is the fitted model, and df is the effective number of parameters or degree of freedom in the model. The effective degree of freedom (df) increases monotonically with N but also depends on the number of basis functions used in the model. Thus, GCV penalizes the addition of basis functions (transformed variables) to the model because it reduces the accuracy of each basis function (i.e., because there are fewer points in each interval). Because MARS tests many possible variables and possible basis functions, the effective degrees of freedom used in parameterizing the model are much higher than the actual number of basis functions in the final model. Steinberg (2001) suggests that k should be two to five times the number of basis function in the model.

The standard mean squared residual error (MSE) is defined as

$$MSE = \frac{\sum_{i=1}^{N}(y_i - \hat{f}(x_i))^2}{N - df}. \quad \text{(Eq. 9.5)}$$

By combining MSE and GCV, we have the following relationship

$$GCV = MSE \frac{N}{(N - df)^2}. \quad \text{(Eq. 9.6)}$$

This relationship shows that GCV will always be larger than MSE. Adding a basis function always reduces the MSE. MARS uses GCV to find the "optimal or best" model that has the lowest GCV measure.

9.3. Setting Control Parameters and Refining Models

MARS allows the modeler to set control parameters to explore the different models and find the "best" model. This section focused on the major parameters available in MARS (Salford Systems, 2001).

Although the default setting for the maximum number of basis functions is 15, MARS allows the user to change this setting if needed. The maximum number of basis functions usually depends on the size of the data set. The larger the data set, the higher the maximum number of basis functions. A rule of thumb is that this maximum number of basis functions should be at least two to four times the number of basis functions in the final model (i.e., the model having the smallest GSCV value). Because that number is usually unknown, an appropriate number should be determined by trial and error. From experience, as the size of consumer data sets is not extremely large, the default number (15) is often sufficient.

Much like principal components, basis functions can be considered as a type of original variable transformation. However, in some cases, the analyst does not want any transformations of selected variables because of a priori knowledge or because of variable interpretability. MARS allows the user to forbid transformations of selected variables into basis functions if needed. If transformations are forbidden on all variables, MARS will produce a variation of a conventional stepwise regression. This model can be used as a baseline from which to measure the benefits of transformation (i.e., basis functions; Salford Systems, 2001).

Although a variable can be forced into a conventional regression model, there is no simple way to force variables into a MARS model. If a variable (Z) has to be represented into a MARS model, residuals $\hat{e} = Y - \hat{Y}$ from linear regression $\hat{Y} = \hat{\beta}_0 + \hat{\beta}_1 Z$ need to be calculated. Then e can be used as new response variable in the MARS model and all other variables (including Z) as candidate predictors. Note that Z needs to be included to capture nonlinearity.

MARS uses the concept of penalty to obtain a parsimonious model and exclude highly correlated variables from the model. The default value for penalty is none, meaning that MARS favors adding new variables to obtain the optimal model. As the penalty is increased, MARS tends to create new knots in existing variables or generate interaction terms involving existing variables, resulting in the optimal model containing few variables. The penalty was originally introduced to deal with multicollinearity. Suppose, for example, that X_1, X_2, and X_3 are all highly correlated. If X_1 is entered into the model first and there is a penalty on added variables, MARS will lean toward using X_1 exclusively instead of some combination of X_1, X_2, and X_3. If the correlation between the variables is quite high, there will be little loss in fit as a result.

By default, MARS generates a knot at every observed scale value so that the slope or direction of the MARS model can be changed anywhere and as often as the data dictate. This could result in overfitting of the data, which is undesirable in many applications. An effective way to restrain knot placement is to specify a minimum number of observations between knots.

By default, interactions are not considered in the MARS model (called the main-effect model). There is no guarantee that a main-effect model adequately fits the data. Two- or three-way interactions can be added into the model (called an interaction model). One can either exclude a variable from any interactions or specify in detail which interactions are allowed and which are forbidden. If the interaction model substantially improves the model fit (i.e., it substantially lowers the GCV value), it, rather than the main-effect model, should be used. If there is not improvement in GCV, interactions should not be included in the model.

MARS also allows us to set the search intensity for the model. The speed parameter is set by default to 4 but can be lowered to 1, 2, or 3 or increased to 5. A speed setting of 1 forces MARS to test every possible knot at every forward step. This ensures that MARS will miss nothing in its model-building phase. It is suitable for small data sets but is extremely slow on large data sets. In contrast, a speed setting of 5 results in a "quick and dirty" estimation. A high-speed setting is usually used for a very large data set. For most applications, the default value (i.e., 4) offers judicious search, high speed, and additional protection from overfitting. However, model results may differ if the speed settings are different.

9.4. Example of MARS Application

The fried mozzarella cheese sticks data set (see section 1.3) is used to illustrate the application of MARS to understanding parameters driving liking for this type of product. The MARS models were fitted using MARS for Windows (Version 2, Salford Systems, San Diego, Calif.). MARS allows the modeler to set control parameters to explore the different models and find the best model. The maximum number of knots was determined by trial and error, two-way interactions were evaluated, and the default values were used for the remaining control parameters (i.e., penalty on added variables, minimum number of observations between knots [minimum span], search speed parameter). Overall acceptance, as well as liking of appearance, flavor, and texture and purchase intent, was modeled using MARS. Predictors used were either hedonic (appearance, flavor, and texture) or diagnostic scales. The results are presented in the following section.

9.4.1. Overall Acceptance

The basis functions and the final MARS models (for the given control parameters, the best model found by MARS is referred to as the final model) for different prespecified maximum numbers of basis functions and interactions are listed in Table 9.1. The table shows that the number of the significantly important basis functions selected by MARS varied between 3 and 10, depending on the prespecified maximum numbers of basis functions and interactions. The various maximum numbers of basis functions and interactions presented in Table 9.1 were chosen only for illustration purposes to demonstrate the role control parameters play on the end modeling results.

If the interactions between the predictor variables are not allowed, the corresponding models are called additive models. In this study, the best additive model would be the one with the smallest GCV value of 0.945 (Table 9.1). This model is the linear combination of the five basis functions (BF1, BF2, BF3, BF4, and BF6), which use the three original hedonic predictors (appearance, flavor, and texture). Functions BF1, BF3, and BF6 increase consumer overall acceptance, whereas functions BF2 and BF4 decrease overall acceptance. In other words, the products with high scores for the acceptance of flavor, appearance, and texture would have high score for overall acceptance, whereas the products with low scores for the acceptance of flavor, appearance, and texture would have low

Table 9.1. Basis functions for predicting overall acceptance from the acceptance of appearance, flavor, and texture (observations $n = 1304$).

Number of Knots	Interactions	Basis Functions	Adjusted R^2	Generalized Cross Validation
3	No	BF1 = max(0, Flavor − 3.0); BF2 = max(0, 3.0 − Flavor); BF3 = max(0, Appearance − 5.0); $Y = 2.516 + 0.583 \times$ BF1 $− 0.753 \times$ BF2 $+ 0.332 \times$ BF3	0.751	1.043
4	No	BF1 = max(0, Flavor − 3.0); BF2 = max(0, 3.0 − Flavor); BF3 = max(0, Appearance − 5.0); BF4 = max(0, 5.0 − Appearance); $Y = 3.967 + 0.590 \times$ BF1 $− 0.664 \times$ BF2 $+ 0.270 \times$ BF3 $− 0.440 \times$ BF4		
5–6	No	BF1 = max(0, Flavor − 3.0); BF2 = max(0, 3.0 − Flavor); BF3 = max(0, Appearance − 5.0); BF4 = max(0, 5.0 − Appearance); BF5 = max(0, Texture − 1.0); $Y = 3.287 + 0.481 \times$ BF1 $− 0.581 \times$ BF2 $+ 0.200 \times$ BF3 $− 0.377 \times$ BF4 $+ 0.225 \times$ BF5	0.776	0.947
7–14	No	BF1 = max(0, Flavor − 3.0); BF2 = max(0, 3.0 − Flavor); BF3 = max(0, Appearance − 5.0); BF4 = max(0, 5.0 − Appearance); BF6 = max(0, Texture − 3.0); $Y = 3.722 + 0.475 \times$ BF1 $− 0.674 \times$ BF2 $+ 0.191 \times$ BF3 $− 0.393 \times$ BF4 $+ 0.242 \times$ BF6	0.776	0.945*
15 or more	No	BF1 = max(0, Flavor − 3.0); BF2 = max(0, 3.0 − Flavor); BF4 = max(0, 5.0 − Appearance); BF6 = max(0, Texture − 3.0); BF14 = max(0, Appearance − 3.0); $Y = 3.403 + 0.473 \times$ BF1 $− 0.704 \times$ BF2 $− 0.268 \times$ BF4 $+ 0.243 \times$ BF6 $+ 0.179 \times$ BF14	0.776	0.946
4	Yes	BF1 = max(0, Flavor − 3.0); BF2 = max(0, 3.0 − Flavor); BF3 = max(0, Appearance − 5.0); BF4 = max(0, 5.0 − Appearance); $Y = 3.967 + 0.590 \times$ BF1 $− 0.664 \times$ BF2 $+ 0.270 \times$ BF3 $− 0.440 \times$ BF4	0.752	1.043
7	Yes	BF1 = max(0, Flavor − 3.0); BF2 = max(0, 3.0 − Flavor); BF3 = max(0, Appearance − 5.0); BF4 = max(0, 5.0 − Appearance); BF6 = max(0, Texture − 1.0); $Y = 4.251 + 0.190 \times$ BF1 $− 0.792 \times$ BF2 $+ 0.179 \times$ BF3 $− 0.469 \times$ BF4 $+ 0.064 \times$ BF5	0.777	0.940
8–9	Yes	BF1 = max(0, Flavor − 3.0); BF2 = max(0, 3.0 − Flavor); BF3 = max(0, Appearance − 5.0); BF4 = max(0, 5.0 − Appearance); BF6 = max(0, 5.0 − Texture) $\times$ BF1; BF7 = max(0, Texture − 1.0); BF8 = max(0, Flavor − 1.0) $\times$ BF4; $Y = 3.411 + 0.546 \times$ BF1 −	0.783	0.920

(*Continued*)

Table 9.1. *Continued.*

Number of Knots	Interactions	Basis Functions	Adjusted R^2	Generalized Cross Validation
		$1.080 \times BF2 + 0.170 \times BF3 - 0.060 \times BF6 + 0.183 \times BF7 - 0.116 \times BF8$		
15	Yes	BF1 = max(0, Flavor − 3.0); BF2 = max(0, 3.0 − Flavor); BF3 = max(0, Appearance − 5.0); BF4 = max(0, 5.0 − Appearance); BF7 = max(0, Texture − 1.0); BF8 = max(0, Flavor − 1.0) × BF4; BF9 = max(0, Texture − 8.0) × BF4; BF10 = max(0, 8.0 − Texture) × BF4; BF11 = max(0, Flavor − 1.0) × BF7; BF12 = max(0, Flavor − 4.0) × BF7; $Y = 3.623 + 0.371 \times BF1 - 0.679 \times BF2 + 0.157 \times BF3 - 0.090 \times BF8 - 0.841 \times BF9 - 0.044 \times BF10 + 0.067 \times BF11 - 0.046 \times BF12$	0.787	0.913

Note: Y = overall acceptance score.
*Best or final model.

scores on overall acceptance. Functions BF1 and BF2 are related to the predictor flavor in a nonlinear fashion approximated by MARS, using two piecewise linear regression lines. Similarly, BF3 and BF4 are the transformed variables for the predictor appearance, and BF5 is for the predictor texture.

BF1 and BF2 are the standard basis function (i.e., *X*-knot) and mirror-image basis function (i.e., knot-*X*), respectively, for the predictor flavor. The graphic representation of the BF1 and BF2 is given in Figure 9.5a. The flavor score of 3 (dislike moderately) is found to be the knot for the predictor flavor. The knot divided the whole space (scale from 1 to nine [dislike extremely to like extremely]) into two local regions: 1–3 (dislike extremely to dislike moderately) and 3–9 (dislike moderately to like extremely). The equation BF1 = max(0, Flavor-3) can be explained as a linear regression line between the response overall acceptance and the predictor flavor only over the region of 3–9 and zero outside of this region. In the same manner, BF2 = max(0, 3-Flavor) linearly related the response to the predictor only over the region of 1–3. Functions BF1 and BF2 are mutually exclusive over the whole region of 1–9. MARS combines BF1 and BF2 into (0.475 × BF1–0.674 × BF2) to describe the relationship between the overall acceptance and the flavor acceptance over the whole region of 1–9. The slope (0.674) for the region of 1–3 is slightly larger than that (0.475) for the region of 3–9 (Figure 9.5a), indicating that 1-unit increase of flavor acceptance scores will increase consumer overall acceptance more for products with very low flavor acceptance (dislike extremely to dislike moderately) than for those with high flavor acceptance (dislike moderately to like extremely). The slope can be interpreted as the increase of the response (such as overall acceptance) by the unit increase of the predictor (such as flavor acceptance).

Similarly, BF3 and BF4 are the standard and mirror-image basis functions, respectively, for the predictor appearance. The knot location is found to be at the appearance score of 5 (Figure 9.5b), which split the whole region of 1–9 into two local regions: 1–5 and 5–9.

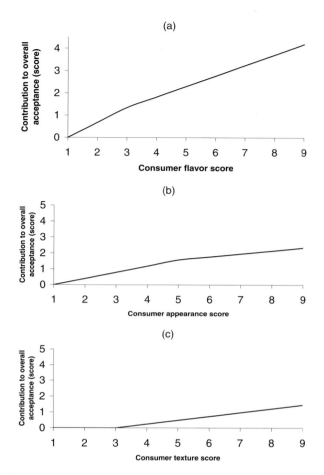

Figure 9.5. Contribution of flavor, appearance, and texture acceptance (see Table 1.9 for hedonic and Just About Right scale definitions) to consumer overall acceptance according to MARS modeling with no interactions.

The equation BF3 = max(0, Appearance-5) is the linear regression line over the local region of 5–9, whereas BF4 = max(0, 5-Appearance) is the linear regression line over the local region of 1–5. The linear combination (0.191 × BF3–0.393 × BF4) of BF3 and BF4 is used to describe the effect of the predictor appearance on the response (i.e., overall liking) over the entire region (1–9). The slopes for the linear regression lines in two local regions of 5–9 and 1–5 are 0.191 and 0.393, respectively. As appearance scores decrease, the overall scores decrease in a piecewise linear regression manner. The drop rate (slope) is larger over the region of 1–5 than over the region of 5–9.

For texture acceptance, there is only one standard basis function, BF6. The knot location is at 3 (Figure 9.5c). The equation BF6 = max(0, Texture-3) is the linear regression line over the local region of 3–9 and zero over the region of 1–3. The effect of the predictor texture on the overall acceptance could be explained using 0.242 × BF6. Figure 9.5c shows that texture has no contribution to the overall acceptance if the texture score is 3 or less. Over the local region of 3–9, the scores of overall acceptance will increase by 0.242 per unit increase of texture acceptance score.

MARS also computes the relative importance value for the predictive variables in the final model in terms of GCV values. For the best additive model, the relative importance values are 100%, 56.80%, and 42.16% for flavor, appearance, and texture, respectively, indicating that the order of the relative contribution to overall acceptance from the most to the least is Flavor > Appearance > Texture. By comparing the contributions for the predictors, the same information can be deduced from Figure 9.5.

MARS uses adjusted R^2 (Adj R^2) and GCV values to assess the goodness of fit of a model. For the final additive model for overall acceptance, Adj R^2 and GCV values are 0.776 and 0.945, respectively. Because MARS treats each product evaluated by each individual consumer as an observation, there are 1304 observations in total for 8 products and 163 consumers in this study. By considering the 1304 observations, an Adj R^2 of 0.776 indicates good fit. Using the final additive model, the mean scores of overall acceptance are predicted, and then the observed and predicted mean scores are compared and presented in Figure 9.6. The R^2 between the observed and predicted mean scores of overall acceptance is 0.987, indicating that the final additive MARS model could predict the mean scores of overall acceptance very well (Figure 9.6).

If the interactions between the predictors are allowed, MARS could generate many more models than without interactions being allowed, and the corresponding models are called interaction models. The selected interaction models are also listed in Table 9.1. Below the maximum number of eight knots (Table 9.1), the interaction models do not improve fit in terms of Adj R^2 and GCV values compared with the additive models (i.e., main-effect models). In contrast, Adj R^2 and GCV values are improved as the maximum number of knots increases from eight to 15 or more. The best interaction model with the smallest GCV value (0.913) is the linear combination of eight basis functions (BF1, BF2, BF3, BF8, BF9, BF10, BF11, and BF12). Functions BF8, BF9, and BF10 use BF4 to express the interactions between appearance and flavor/texture. In the same way, BF11 and BF12 use BF7 to express the interactions between flavor and texture. The interactions between the predictive variables are presented in Figure 9.7. By examining Figure 9.7c, it is clear that at the appearance acceptance score of 1, the relative contribution to overall acceptance is 2.132 for a texture acceptance score of 1, whereas it is 0.0 at the texture acceptance score of 9. This is not logical and reasonable, indicating the possible

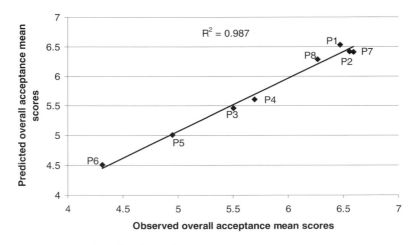

Figure 9.6. Observed and predicted overall acceptance mean scores.

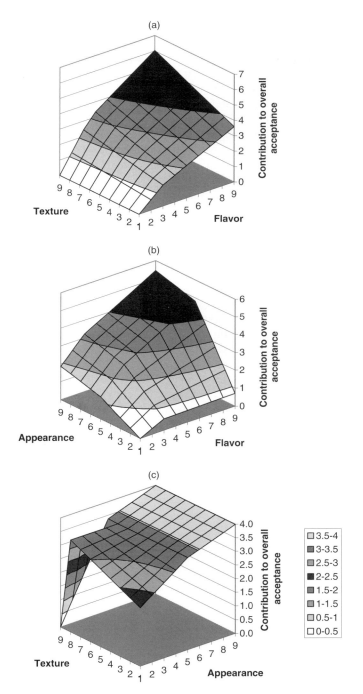

Figure 9.7. Contributions of flavor, appearance, texture, and interactions (see Table 1.9 for hedonic and Just About Right scale definitions) to consumer overall acceptance according to MARS modeling with interactions.

overfitting because 10 (i.e., too many) basis functions are used. When more basis functions are used and more interactions are allowed, MARS will create more knots, which divide the predictor space into a greater number of smaller local regions, and then MARS will fit the observed data in each local region more accurately, regardless of noise or signal data. This overfitting is usually undesirable in many applications, and a smoother—albeit less locally accurate—model might be preferred (Salford Systems, 2001).

Using the final interaction model, the mean scores of overall acceptance were predicted, and the R^2 between the observed and predicted mean scores of overall acceptance is 0.985. Although the final interaction model uses five more basis functions than the final additive model, its R^2 value (0.985) does not increase compared with the R^2 (0.987) for the final additive model. In addition, the predicted overall acceptance mean scores obtained from both models are almost identical. Friedman (1991) suggests that the interaction model is chosen over the additive model only if its Adj R^2 is substantially larger. In this case, we determine that the additive model is better to model overall acceptance and that interactions between predictive variables do not have to be considered.

9.4.2. Flavor Acceptance

In this study, only one flavor attribute (saltiness) was evaluated on the JAR scale by the consumers. It has been found in previous research that it can be difficult to relate the JAR predictive variables (such as JAR saltiness, JAR size, JAR color, etc.) to a response variable such as consumer acceptance evaluated on a 9-point hedonic scale. The conventional linear regression models are not appropriate for JAR scale data because the best or ideal score is in the middle of the JAR scale. As the scores of the JAR predictor (saltiness) are far away from the JAR score (3 for a 5-point JAR scale) on both sides (3–1 and 3–5), consumer acceptance scores would be expected to either drop if salt has an effect on liking or stay constant if not. It is also possible that the drop rate on both sides of the JAR score may differ or even vary over the JAR scale, which conventional linear regression models cannot handle. MARS provides promising results to solve this problem (Table 9.2 and Figure 9.8). Two standard basis functions, BF2 and BF3, are found to be significant ($p < 0.05$). Equation BF2 = max(0, 3.0-Saltiness) has a knot location at 3 that is the JAR score, which splits the whole scale into two local regions. Similar to BF2, BF3 = max(0, Saltiness-2.0) divides the region at the JAR score of 2 into two subregions.

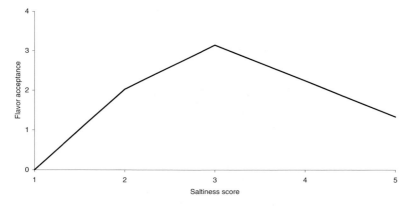

Figure 9.8. Contribution of saltiness (see Table 1.9 for hedonic and Just About Right scale definitions) to flavor acceptance according to MARS modeling.

Table 9.2. Basis functions for predicting flavor, appearance, and texture acceptance and purchase intent from diagnostic attributes (observations n = 1304).

Number of Knots	Interactions	Basis Functions	Adjusted R^2	Generalized Cross Validation
4–40	No	Flavor acceptance: BF2 = max(0, 3.0 − Saltiness); BF3 = max(0, Saltiness − 2.0); Y = 7.212 − 2.026 × BF2 − 0.907 × BF3	0.122	3.954*
15	No	Appearance acceptance: BF1 = max(0, Color − 3.0); BF2 = max(0, 3.0 − Color); BF3 = max(0, Size − 3.0); BF4 = max(0, 3.0 − Size); BF5 = max(0, Breading − 3.0); BF6 = max(0, 3.0 − Breading); BF7 = max(0, Size − 2.0); BF9 = max(0, Size − 4.0); Y = 6.823 − 1.938 × BF1 − 1.216 × BF2 − 1.112 × BF3 − 0.582 × BF4 − 0.836 × BF5 − 0.668 × BF6 + 0.631 × BF7 − 2.122 × BF9	0.491	2.314*
4–40	No	Texture acceptance: BF1 = max(0, Crispiness − 3.0); BF2 = max(0, 3.0 − Crispiness); BF3 = max(0, CheeseTexture − 3.0); BF4 = max(0, 3.0 − CheeseTexture); Y = 7.300 − 1.667 × BF1 − 1.187 × BF2 − 1.155 × BF3 − 1.299 × BF4	0.448	2.310*
14	No	Purchase intent acceptance: BF1 = max(0, Overall − 5.0); BF2 = max(0, 5.0 − Overall); BF5 = max(0, Flavor − 1.0); BF7 = max(0, 3.0 − Crispiness); BF8 = max(0, CheeseTexture − 3.0); BF9 = max(0, 3.0 − CheeseTexture); BF10 = max(0, Overall − 8.0); BF12 = max(0, Breading − 3.0); BF13 = max(0, 3.0 − Breading); BF14 = max(0, Texture − 4.0); Y = 1.617 + 0.375 × BF1 − 0.066 × BF2 + 0.143 × BF5 − 0.164 × BF7 − 0.159 × BF8 − 0.126 × BF9 − 0.387 × BF10 − 0.131 × BF12 − 0.109 × BF13 + 0.134 × BF14	0.651	0.61*
15	Yes	Appearance acceptance: BF1 = max(0, Color − 3.0); BF2 = max(0, 3.0 − Color); BF3 = max(0, Size − 3.0); BF4 = max(0, 3.0 − Size); BF5 = max(0, Breading − 3.0); BF6 = max(0, 3.0 − Breading); ABF7 = max(0, Size − 1.0) × BF2; BF12 = max(0, Breading − 1.0); Y = 7.486 − 3.105 × BF1 − 0.777 × BF2 − 0.503 × BF3 − 1.234 × BF4 − 0.914 × BF5 − 0.614 × BF6 − 0.328 × BF7 + 0.539 × BF9	0.495	2.329
5	Yes	Texture acceptance: BF1 = max(0, Crispiness − 3.0); BF2 = max(0, 3.0 − Crispiness); BF3 = max(0, CheeseTexture − 3.0); BF4 = max(0, 3.0 − CheeseTexture); BF6 = max(0, 3.0 − CheeseTexture) × BF2; BF7 = max(0,	0.459	2.307

(*Continued*)

Table 9.2. Continued.

Number of Knots	Interactions	Basis Functions	Adjusted R^2	Generalized Cross Validation
		CheeseTexture − 4.0); BF9 = max(0, Crispiness − 3.0) × BF7; BF11 = max(0, Crispiness − 2.0) × BF7; BF15 = max(0, Crispiness − 1.0) × BF4; Y = 7.328 − 1.756 × BF1 − 1.283 × BF2 − 0.971 × BF3 − 0.576 × BF6 + 2.761 × BF9 − 1.362 × BF11 − 0.681 × BF15		
8–9	Yes	Purchase intent acceptance: BF1 = max(0, Overall − 5.0); BF2 = max(0, 3.0 − Overall); BF3 = max(0, Texture − 2.0); BF4 = max(0, 2.0 − Texture); BF6 = max(0, 8.0 − Flavor) × BF3; BF8 = max(0, 3.0 − Crispiness); BF9 = max(0, CheeseTexture − 3.0) × BF1; BF10 = max(0, 3.0 − CheeseTexture) × BF1; BF13 = max(0, Breading − 3.0) × BF1; BF14 = max(0, 3.0 − Breading) × BF1; BF15 = max(0, Breading − 4.0) × BF4; BF17 = max(0, Texture − 6.0) × BF8; BF18 = max(0, 6.0 − Texture) × BF8; BF19 = max(0, Texture − 6.0); BF20 = max(0, 6.0 − Texture); BF21 = max(0, Appearance − 1.0) × BF20; Y = 1.251 + 0.400 × BF1 − 0.147 × BF2 + 0.447 × BF3 − 0.039 × BF6 − 0.322 × BF8 − 0.122 × BF9 − 0.119 × BF10 − 0.113 × BF13 − 0.098 × BF14 + 0.1341 × BF15 + 0.147 × BF17 + 0.089 × BF18 − 0.398 × BF19 + 0,020 × BF21	0.661	0.607

Note: Y = overall acceptance score.
*Best or final model.

Figure 9.8 shows the linear combination of these two basis functions; namely, flavor acceptance score $Y = 7.212 − 2.026 \times BF2 − 0.907 \times BF3$. As the saltiness scores stray away from the JAR score, flavor acceptance scores decrease. The drop rate is faster over the local region of 1–2 than over the local region of 4–5. This indicates that too little salt is more detrimental to consumer flavor acceptance than too much salt. The Adj R^2 (0.122, equivalent $R = 0.35$) is lower, probably because of an inability to capture all the drivers of flavor liking from just the salt variable. The variation among the eight products and 163 consumers (1304 observations) also lowers the Adj R^2 value. However, the R^2 between the observed and predicted mean scores of flavor acceptance is 0.87, indicating that the fitted MARS model could predict the mean score of flavor acceptance reasonably well.

9.4.3. Appearance Acceptance

Given the maximum numbers of knots and interactions allowed, the best models selected by MARS are provided in Table 9.2. Without considering interactions between the predictors, the MARS model ($p < 0.05$; Table 9.2) relates the consumer appearance acceptance scores to three predictive variables evaluated on a JAR scale (color, size, and breading), with an Adj R^2 of 0.486 ($R = 0.70$). As discussed previously, the lower Adj R^2 is a result of the small number of predictors and the variation among 1304 observations. The fitted model is used to calculate the predicted means of appearance acceptance scores, and the R^2 value between the observed and predicted means for appearance acceptance scores is 0.93, indicating that the MARS model captured the relationship between liking and diagnostic attributes.

BF1 and BF2 are the standard and mirror-image basis functions, respectively, for the product's color. The JAR score of 3 is found to be the knot location (Figure 9.9a). The

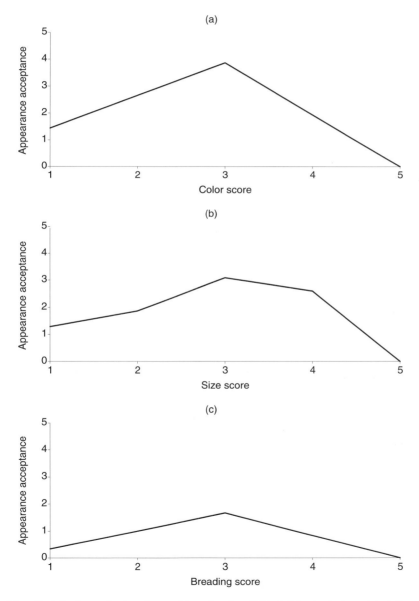

Figure 9.9. Contributions of color, size, and breading (see Table 1.9 for hedonic and Just About Right scale definitions) to appearance acceptance according to MARS modeling.

linear combination of BF1 and BF2, namely, $-1.938 \times$ BF1 $- 1.216 \times$ BF2, describes the effect of the color on appearance acceptance. The equation $-1.938 \times$ BF1 $= -1.938 \times$ max(Color-3) means that appearance acceptance scores decrease by 1.938 per unit increase of color score over the region of 3–5. Similarly, $-1.216 \times$ BF2 $= -1.216 \times$ max(3-Color) means that a 1-unit drop of color score in the range of 1–3 would decrease the appearance acceptance score by 1.216. The coefficient of 1.938 for BF1 is greater than the coefficient of 1.216 for BF2, which indicates that too dark of a color (4–5) has a more detrimental effect on appearance acceptance than too light a color (2–1).

Size has four basis functions and three knots (Table 9.2). The equations BF3 = max(0, Size-3) and BF4 = max(0, 3-size) are the paired standard and mirror-image basis functions at the JAR score of 3, whereas BF7 = max(0, Size-2) and BF9 = max(0, Size-4) are the standard basis functions at the knots of 2 and 4, respectively. The piece-wise linear regression (-0.112BF3 $- 0.582$BF4 $+ 0.631$BF7 $- 2.122$BF9) describes the contribution of product size to appearance acceptance. Figure 9.9b clearly shows that the four local regions have different slopes, indicating that the size effects vary from region to region on the JAR scale. The JAR size has the highest appearance acceptance, followed by the size score between 3 and 4 (slope = $-1.112 + 0.631 = -0.481$), the size score between 3 and 2 (slope = $0.582 + 0.631 = 1.215$), the size score between 2 and 1 (slope = 0.582), and the size score between 4 and 5 ($-1.112 + 0.631 - 2.122 = -2.603$). This implies that a "somewhat too large" size has a less negative effect on acceptance than does a "somewhat too small" size, whereas a "much too large" size has a more negative effect on acceptance than does a "much too small" size. The attribute breading shows similar results to color (Figure 9.9c). The linear combination ($-0.836 \times$ BF5 $+ 0.668 \times$ BF6) in the model indicates that consumers were less bothered by not enough breading than by too much breading because the drop rate of 0.668 over the region of 3–1 is smaller than the drop rate of 0.836 over the region of 3–5. The relative variable importance values are 100.00%, 70.93%, and 52.25% for the color, size, and breading, respectively, indicating that color is the most important contributing factor to appearance acceptance.

9.4.4. Texture Acceptance

The MARS models for texture acceptance are presented in Table 9.2 and Figure 9.10. Four basis functions are given by MARS—a pair (BF1 and BF2) for crispiness and a pair (BF3 and BF4) for cheese texture. The regression coefficients of -1.667 for BF1 = max(0, Crispiness-3) and -1.187 for BF2 = max(0, 3-Crispiness) mean that the drop rate is faster over the region of 3–5 than over the region of 3–1, indicating that "too crispy" is more detrimental to the acceptance of texture than "not crispy enough." The regression coefficients of -1.155 for BF3 = max(0, CheeseTexture-3) and -1.299 for BF4 = max(0, 3-CheeseTexture) mean that the drop rate is slower over the region of 3–5 than over the region of 3–1, implying that "too soft/melted" is more detrimental to texture acceptance than "too firm/not melted." In terms of MARS relative variable importance values (100% and 96.73%, respectively), crispness seemed to be slightly more important to the texture acceptance than cheese texture. The R^2 between the observed and predicted mean scores of texture acceptance is 0.95, indicating that the fitted MARS model could accurately predict the mean scores for texture acceptance. The use of interactions in the model did not improve the R^2 and GCV values. Therefore, a discussion for the interaction model is not given here.

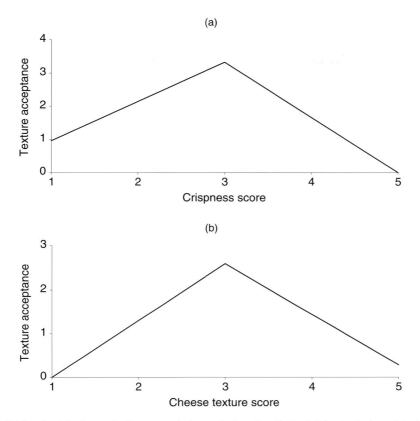

Figure 9.10. Contributions of crispness and cheese texture (see Table 1.9 for hedonic and Just about Right scale definitions) to texture acceptance based on MARS modeling.

9.4.5. Purchase Intent

The MARS modeling results for consumer's purchase intent are presented in Table 9.2 and Figure 9.11. Because the interaction model does not offer improvement of the fit (Table 9.2), only the main-effect model is discussed here.

Three standard basis functions (BF1, BF2 and BF3) for overall acceptance have two knots located at 5 and 8 on the 9-point hedonic scale (Figure 9.11a). The small slope (0.066 in Table 9.2) for the linear line between 1 (dislike extremely) and 5 (neither like nor dislike) means that the contribution of overall acceptance to purchase intent is rather small, which seems to be quite reasonable because consumers usually do not buy products they dislike. As liking increases from 5 to 8, consumers' purchase intent increases by 0.375 per unit score of overall acceptance. Above 8 points, purchase intent does not change as the score of overall acceptance increases further.

Although MARS gives only one basis function to flavor acceptance, flavor virtually has no transformation (basis function). The equation BF5 = max(Flavor-1.0) describes a linear relation between flavor acceptance and purchase intent. At the flavor acceptance score of 9, flavor has the maximum contribution (0.143max(Flavor-1) = 0.143 × 8 = 1.144) to purchase intent. As the scores decrease from 9 down to 1, the contribution also decreases at the rate of 0.143 per unit of flavor acceptance.

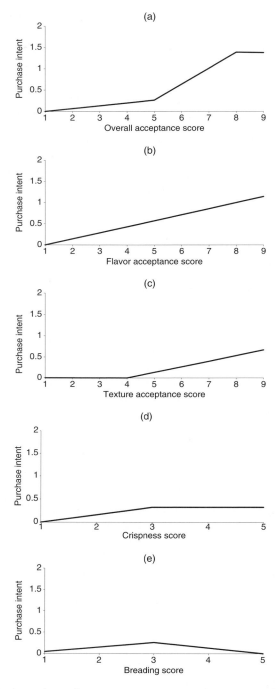

Figure 9.11. Contributions of overall acceptance, acceptance of appearance, flavor, texture, amount of breading, and crispiness to purchase intent based on MARS modeling. See Table 1.9 for hedonic and Just About Right Scale designations.

The only knot found by MARS for texture acceptance basis functions is 4. This indicates that below 4 (dislike extremely to dislike slightly), texture acceptance does not affect a consumer's purchase intent. As the score of texture acceptance increases from 4 to 9, texture acceptance makes a contribution to purchase intent of 0.134 per unit increase of texture acceptance score.

The basis function BF7 = max(0, 3-Crispiness) shows that purchase intent score increases as crispiness score increases up to JAR score of 3, and above the JAR score, purchase intent remains constant. This implies that "too crispy" does not increase or decrease purchase intent, whereas "not crispy enough" decreases purchase intent. Breading has similar effects on purchase intent (Figure 9.11e), and their contributions are relatively small. The relative importance scores for overall acceptance, acceptance of appearance, flavor, texture, JAR crispiness, JAR breading, and JAR cheese texture are 100.0%, 65.0%, 50.8%, 35.1%, 26.6%, and 3.0%, respectively. Overall and flavor acceptance have the most contribution to purchase intent.

9.5. A Comparison with PLS Regression

Linear and quadratic PLS regression models are alternative modeling techniques for studying the relationship between hedonic scores and attribute intensities or appropriateness measures derived from consumer testing. To compare MARS and PLS models, the same cheese sticks data set is used here, and similar analyses to those performed with MARS are offered in the PLS regression framework.

9.5.1. Overall Acceptance

The consumer overall acceptance is modeled using consumer appearance, flavor, and texture scores as predictors. The R^2 and RMSEP (root mean square error of prediction) for the predictive PLS model ($n = 1304$) are 0.77 and 0.92, respectively, indicating that the PLS model fitted well. Figure 9.12 is a graphic representation of observed versus predicted consumer overall acceptance mean scores. It is evident from the figure that P1, P7, P2, and P8 have the highest overall acceptance, whereas P6 has the lowest acceptance. The mean scores are predicted well by the PLS model and are almost identical to those predicted by

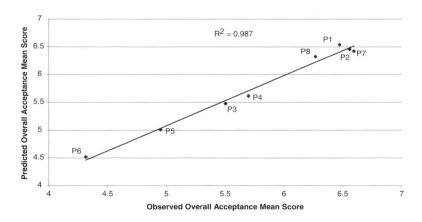

Figure 9.12. Observed and predicted overall acceptance mean scores (partial least squares model).

Table 9.3. R^2, RMSEP, and weighted regression coefficients for quadratic PLS models (number of observations $n = 1304$).

Response	Important Predictors	Weighted Regression Coefficients		Partial Least Squares Model Root Mean Square Error Of Prediction	
		Linear	Quadratic	R^2	Prediction
Overall liking = f (attribute liking)				0.76	0.99
Overall acceptance	Appearance acceptance	0.141	0.01		
	Flavor acceptance	0.267	0.022		
	Texture acceptance	0.127	0.009		
Appearance liking = f (size, breading, color)				0.21	1.85
Appearance acceptance	Size (JAR)	2.87	−0.439		
	Breading (JAR)	3.041	−0.523		
	Color (JAR)	4.208	−0.741		
Flavor liking = f(salt)				0.02	2.1
Flavor acceptance	Salt (JAR)	4.498	−0.721		
Texture liking = f (crispness, cheese texture)				0.41	1.55
Texture acceptance	Crispness (JAR)	4.099	−0.696		
	Cheese texture (JAR)	3.387	−0.567		

Note: JAR = Just about Right.

the MARS model. For the PLS model, the weighted regression coefficients are used to assess the relative importance of each predictive variable to the prediction of the response variable (Meullenet et al., 2001). Table 9.3 shows the weighted regression coefficients for the PLS model predicting consumer overall acceptance from the acceptance scores for appearance, flavor, and texture. The results from the jackknife optimization shows that all three attributes are important (i.e., significant) in determining consumer overall acceptance, with flavor being the most important, followed by texture and appearance. The order of the relative contribution of flavor, appearance, and texture is the same for both PLS and MARS models.

9.5.2. Appearance Acceptance

Consumer appearance acceptance is modeled using JAR size, JAR breading, and JAR color as predictors. In this analysis, the quadratic terms of the predictors are included to attempt the modeling of nonlinear relationships between hedonic and JAR responses. All six predictive variables were standardized before PLS (i.e., centered and weighted by the reciprocal of their respective standard deviations). This was done so that predictive variables with large standard deviations or magnitude (i.e., the quadratic terms) were not allowed to dominate the model. All three JAR variables are important to the acceptance of appearance according to the jackknife method that can be used in PLSR software such as Unscrambler. The R^2 and RMSEP for the PLS model ($n = 1304$) are 0.42 and 1.58, respectively. The R^2 value between the observed and predicted means of appearance acceptance scores is 0.86, indicating that the PLS model could make adequate predictions of mean appearance acceptance scores. In comparison to the MARS model, the PLS model

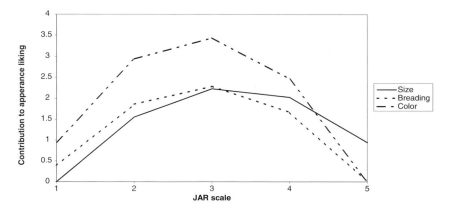

Figure 9.13. Contributions of color, size, and amount of breading to appearance acceptance (see Table 1.9 for hedonic and Just about Right scale definitions) from quadratic partial least squares regression modeling.

yielded a lower R^2 value, indicating that MARS was better for modeling the relationship between appearance liking and JAR variables. As seen in Figure 9.13, size was found to be the most important contributor to appearance, which is a similar finding to that of MARS. However, the contribution of the amount of breading and color was similar in the PLS model, whereas Size was more important according to MARS. Quadratic terms are useful when investigating the relationship between liking and JAR data. However, the inclusion of quadratic terms in the PLS model hinders the interpretability of the regression coefficients, though, the model still can be used to establish the role of JAR variables with relative ease (Figure 9.13).

9.5.3. Flavor Acceptance

Only one JAR variable, saltiness (including a quadratic term), is used to predict the consumer flavor acceptance. The R^2 and RMSEP for the PLS model ($n = 1304$) are 0.02 and 2.10, respectively, whereas the R^2 value between the observed and predicted means of flavor acceptance scores is 0.46. The lower R^2 values indicate poor fitting for flavor acceptance. In comparison with the MARS model, the PLS model has a much lower R^2 value. However, Figure 9.14 for the quadratic PLS model is very similar to the results obtained for MARS (Figure 9.8). Therefore, the interpretation of the results is in this case the same for PLS and MARS.

9.5.4. Texture Acceptance

Consumer texture acceptance is modeled here using both JAR crispness and JAR cheese texture as predictors and their quadratic terms. According to the jackknife procedure, both crispness and cheese texture are important predictors of texture liking. The R^2 and RMSEP for the PLS model ($n = 1304$) are 0.41 and 1.55, respectively. The R^2 value between the observed and predicted means of texture acceptance scores is 0.90, indicating that the PLS model could adequately predict texture acceptance mean scores. The PLS model has a slightly lower R^2 value than the MARS model previously described (Table 9.2), indicating that the MARS model was better at modeling the relationship. Figure 9.15 shows the

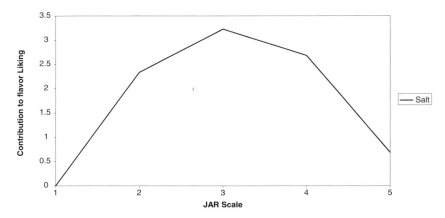

Figure 9.14. Contribution of saltiness to flavor acceptance (see Table 1.9 for hedonic and Just about Right scale definitions) from quadratic partial least squares regression modeling.

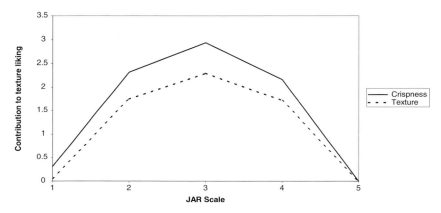

Figure 9.15. Contribution of crispness and cheese texture to texture acceptance (see Table 1.9 for hedonic and Just About Right scale definitions) from quadratic partial least squares regression modeling.

relationship between JAR scores and texture liking. Similar to the MARS model, PLS established that crispness was a slightly more important contributor to texture liking than cheese texture, which is in agreement with the MARS results (Figure 9.10). The main difference between the two models is that MARS determined the relationship in the ranges of 1–3 and 3–5 to be linear, which is not the case for the PLS model. The different slopes reported for PLS are a result of the relationship being forced to be of the quadratic form even if the data would be fitted by two straight lines, as is the case in the MARS model.

In summary, the advantages of the novel modeling technique MARS is that it automates both the selection of variables and the nonparametric transformation of variables (basis functions) to achieve the best fit model and variable transformation is accomplished implicitly through the piecewise linear regression function to express underlying nonlinear functions. The example given shows that MARS has potential value in modeling consumer acceptance of food, especially in exploring complex relationships between response and predictors and relating predictors (such as crispness, saltiness) evaluated on a JAR scale to a response such as consumer acceptance. The use of MARS to model JAR data can provide very useful information for food product developers and is discussed further in Chapter 10. Comparison of MARS with commonly used PLS regression and proportional odds model is ongoing. Further research needs to extend MARS applications into the external preference mapping framework.

References

Craven, P. and G. Wahba. 1979. Smoothing noisy data with spline functions: Estimating the correct degree of smoothing by the method of generalized cross-validation. Numer. Math. 31:377–403.

Dwinnell, W. 2000. Exploring MARS: An alternative to neural networks. PCAI 14(1):21.

Francis, L. 2003. Martian Chronicles: Is Mars better than neural networks, White Paper, Salford Systems. Available at http://www.salfordsystems.com/doc/03wf027.pdf. Accessed January 8, 2007.

Friedman, J. 1991. Multivariate Adaptive Regression Splines. Ann. Stat. 19:1–141.

Golub, G., M. Heath, and G. Wahba. 1979. Generalized cross-validation as a method for choosing a good ridge parameter. Technometrics 21:215–223.

Meullenet, J.-F., V.K. Griffin, K. Carson, G. Davis, S. Davis, J. Gross, J.A. Hankins, E. Sailer, C. Sitakalin, S. Suwansri, and A.L. Vasquez Caicedo. 2001. Rice external preference mapping for Asian consumers living in the United States. J Sensory Stud. 16:73–94.

Salford Systems. 2001. MARS User's Guide. Salford Systems, 2001.

Sephton, P. 2001. Forecasting recessions: can we do better on MARS™? Federal Reserve Bank of St. Louis Review, March/April 2001, 83(2):39–49.

Steinberg, D. 2001. An alternative to neural networks: Multivariate adaptive regression splines. PC AI January/February:38–41.

Xiong, R. and J.-F. Meullenet. 2004. Application of multivariate adaptive regression splines (MARS) to the preference mapping of cheese sticks. J. Food Sci. 69(4):SNQ131–139.

10 Analysis of Just About Right Data

10.1. Introduction

Just About Right (JAR) or diagnostic scales are used extensively in consumer research, primarily to assist product developers with identifying potential shortcomings of the products tested. These scales usually feature from five to seven categories, are balanced, and feature a center category, albeit the Just About Right category (Table 10.1). Individual samples can be rated on a JAR scale as "too weak," "too strong," or "just about right" along a particular continuum. These scales are used to assess the appropriateness of specific sensory levels. This is an important concept, as the liking of a sensory modality is related to its intensity. When the relationship is known to not be linear, it is often parabolic, where the liking for a product will first increase with increasing levels of the attribute, peak, and then decrease as the intensity of the attribute continues to increase. The peak has been described as the bliss point (i.e., the JAR level), and the shape of the curve, known as the inverted U or L function, relates hedonic response to the stimulus intensity first proposed by Wundt (Moskowitz and Jacobs, 1987). It should be noted that the relationships between the sensory experience and the hedonic response and the physical stimulus are governed by very different laws, as the relationship between perceived intensity and stimulus intensity can usually be modelled by a power law (Stevens, 1975).

JAR data that are normally distributed around the center of the scale (i.e., the "just about right" category) are indicative of an optimized level of the continuum or attribute. It is important with this type of data to examine the distribution of the raw data. For example, a consumer panel might contain one segment that prefers products with one set of sensory characteristics and another segment that prefers the same product with a completely different set of sensory characteristics. This is illustrated in Figure 10.1 which is a distribution of scores on a JAR scale for the crispness of cheese sticks.

In this example, the appropriateness of the level of crispness in a fried cheese stick appetizer was accessed with a panel made up of 180 individuals. The distribution of the data clearly shows that they are not distributed around the just right score, although the mean value of scores would be close to the JAR point. Instead, the distribution of the data is bimodal, with roughly equal numbers of consumers who found the product to be either too soggy or too crispy. The appropriateness of the cheese texture for the same product was evaluated by the same consumer panel, and the scores' distribution is given in Figure 10.2. For cheese texture, there was a much greater consensus among consumers as to the appropriateness of the level of cheese firmness. This is in sharp contrast to the results from Figure 10.1. These two examples illustrate the importance of doing a consumer data distribution assessment before more advanced analyses are performed. This is especially true when means are to be used for further analyses such as special cases in which JAR data are used to identify the attributes driving liking.

Although some of the analyses described in this chapter are not all statistical, there is a lack of information in the literature on the analysis of diagnostic scale data. This chapter

Table 10.1. Diagnostic attribute scales used to assess the appropriateness of various attributes of nutritional beverages.

Attribute			Verbal Labels to Categories		
Color	Much too light	Too light	Just about right	Too dark	Much too dark
Chocolate	Much too weak	Too weak	Just about right	Too strong	Much too strong
Sweetness	Not nearly sweet enough	Not sweet enough	Just about right	Too sweet	Much too sweet
Tartness	Not nearly tart/sour enough	Not tart/sour enough	Just about right	Too tart/sour	Much too tart/sour
Thickness	Much too thin	Too thin	Just about right	Too thick	Much too thick
Smoothness	Much too coarse/chalky	Too coarse/chalky	Just about right	Too smooth	Much too smooth

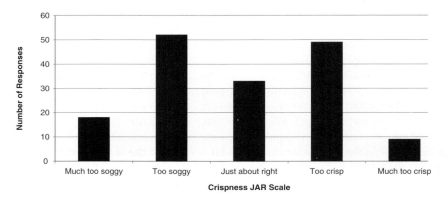

Figure 10.1. Distribution of the JAR scores for crispness of a fried cheese stick appetizer.

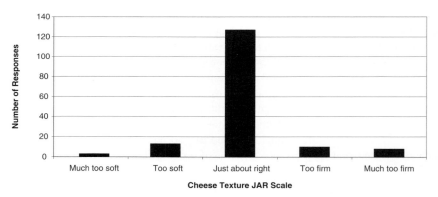

Figure 10.2. Distribution of JAR scores for the cheese texture of a fried cheese stick appetizer.

fills some of these gaps and presents some new applications of multivariate statistical methods as well.

10.2. Basics of Penalty Analysis

Penalty analysis is widely used in the food industry but is not a very well described method in the literature. Although penalty analysis is not a multivariate analysis method or even a statistical method and would seem to be out of the realm of this text, improvements to this technique involving multivariate analysis techniques are discussed later in this chapter, and it seems important to describe the principle of this analysis. The intent of penalty analysis is to determine the effect of specific attributes being at a level less than optimal (i.e., scores above and below the JAR score) on the hedonic level (i.e., liking) of a product. The principle for calculating penalties is given in Figure 10.3. For any specific attribute, consumers are usually divided into three groups (i.e., those consumers who found the attribute to be JAR, those who found the intensity of the attribute to be too weak, and those who found the intensity of the attribute to be too strong).

Hedonic scores for the three groups are calculated as well as the percentage of consumers represented in each of the groups. To conclude that a specific attribute is at its optimal level, a minimum of 70% of the responses are usually expected to be in the JAR group. To conclude that an attribute is not at its optimal level, a minimum of 20% of consumers is usually needed in the "too weak" or "too strong" categories. These are obviously rules of thumb, and the percentages can be adjusted to the user's liking. The penalties are calculated by subtracting the mean hedonic score for the JAR group to either the means of the "too weak" or "too strong" groups. The penalties are usually represented graphically for individual products by scatter plotting the percentage of consumers in a non-JAR group on the x-axis and the penalty on the y-axis. An example of such a plot is given in Figure 10.4. The interpretation is trivial, as the most glaring weaknesses of a product are those with high penalties and a high percentage of consumers finding the attribute to not be at a JAR level. The example given in Figure 10.4 shows that the soup sampled was far from optimized from the sensory standpoint. The overall flavor impact was found to be too strong by over 40% of the respondents and was responsible for a large drop in hedonic scores (>3 points on the 9-point hedonic scale). The size of the vegetable chunks (SIZEJAR) was found to be too small by 35% of the consumers but had a much lower penalty (~1). This type of analysis allows the product developer or sensory analyst to make decisions as to what sensory properties should be dealt with (i.e., improved). Usually, attributes located in the upper-right quadrant of the chart would be selected as those needing to be "fixed." We should also point out that penalty analysis is not without weaknesses. For

Hedonic scores (Y)	Jar scores
7	3
6	4
7	5
8	3
9	3
6	1
5	2
2	4
4	5
6	4
6	1

$\overline{Y}_{<JAR} = 5.7$ Penalty$_{<JAR}$=8.0-5.7=2.3   % consumers$_{<JAR}$=3/11=27%

$\overline{Y}_{JAR} = 8.0$ % consumers$_{JAR}$=28%

$\overline{Y}_{>JAR} = 5.0$ Penalty$_{>JAR}$=8.0-5.0=3.0   % consumers$_{>JAR}$=5/11=45%

Figure 10.3. Penalty analysis calculation principles.

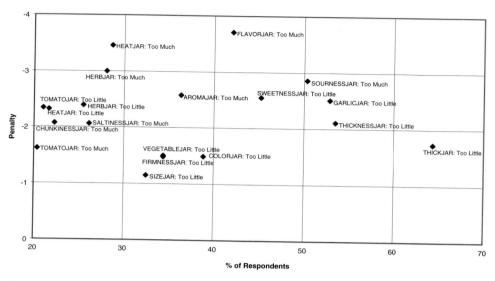

Figure 10.4. Example of penalty analysis for a soup sample.

example, the penalties are obviously not additive, as the sum of the penalties in Figure 10.4 is 42.85 of the 9-point hedonic scale. This stems from the facts that the collinearity among the variables is ignored and that halo effects are a well-known bias in consumer testing. A t-test of the overall hedonic scores comparing scores of a non-JAR and a JAR group for samples of unequal sizes and variance can be employed to test the significance of the penalties. However, with a small number of responses, this test may not be completely satisfactory. We present here an alternative based on resampling techniques.

10.3. Bootstrapping Penalty Analysis

The bootstrap method, developed by Efron in the 1970s, is a computer-based Monte Carlo simulation technique for estimating standard errors, confidence intervals, biases, and prediction errors (Efron, 1982). It has become very popular and is easy to apply. The process of estimating the standard error of a mean drop is illustrated in Figure 10.5. An original data set partially disclosed in Figure 10.5 contains n pairs of overall liking scores and JAR flavor scores. Mean drops for Too Little and Too Much categories were calculated according to penalty analysis principles given in Section 10.2. The bootstrap estimate of variability is obtained through resampling of the data pairs (i.e., liking score, JAR score). Imagine that the data set is replicated an enormous number of times (called bootstrap [B] replications; $B = 10,000$ in Figure 10.5). For each replication, a sample of size n (bootstrap sample) is then selected at random from the original data set. This is, in effect, sampling with replacement, where a data pair is sampled and then returned to the data set and has $1/n$ probability of being drawn again or resampled. A particular pair (Figure 10.5) appears in several of the bootstrap samples and may appear more than once in a particular sample. Each of the B bootstrap samples is used to calculate the mean drops for the Too Little and Too Much categories according to the standard

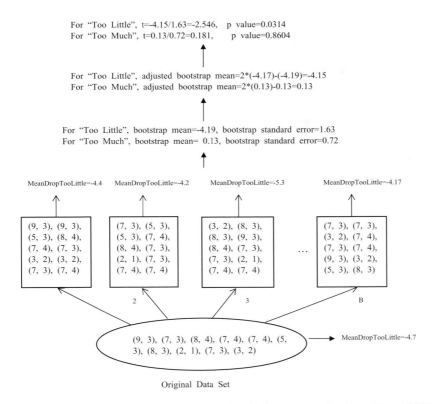

Figure 10.5. Scheme for bootstrapping penalty analysis with bootstrap replications of $B = 10{,}000$.

penalty analysis methodology. On the basis of the B replications of the mean drops, the mean and standard error of the penalties can be computed using Equations (10.1) and (10.2), respectively:

Bootstrap estimate of mean:

$$\bar{s}_b = \frac{\sum_{i=1}^{B} s_i^*}{B}; \quad \text{(Eq. 10.1)}$$

Bootstrap estimate of standard error:

$$\hat{se}_b = \sqrt{\frac{\sum (s_i^* - \bar{s}_b)^2}{B-1}}, \quad \text{(Eq. 10.2)}$$

where s_i^* is the mean drop for the ith bootstrap sample, B is the number of bootstrap samples or replications, $\bar{s}_b$ is the bootstrap estimate of the mean, and $\hat{se}_b$ is the standard error of the mean. Because the bootstrap estimate of the mean is somewhat biased, the bias must be removed using Equation (10.3).

$$\bar{s}'_b = 2s_n - \bar{s}_b, \quad \text{(Eq. 10.3)}$$

where s_n is the estimate of the mean from the original data set and $\bar{s}'_b$ is the adjusted bootstrap estimate of the mean. The number of bootstrap samples (B) is at least 100 and often several thousands. This variable is selected so that the distribution of mean drops is smooth or the variance has stabilized. From the standard error and bootstrap estimate of the penalty, a t-test outlined in Figure 10.5 can easily be performed. The calculated $t = \bar{s}_b/s\hat{e}_b$ can then be compared to a table $t_{\alpha\infty}$.

To apply bootstrapping penalty analysis, one needs to either code one's own program or use the bootstrapping penalty analysis program developed by the authors, which runs in a Microsoft Excel environment. This bootstrapping penalty analysis program is available from the author's Web site for this text (http://www.uark.edu/ua/multivariate/). If the authors' bootstrapping penalty analysis program is used, the steps outlined in Appendix 10.1 need to be followed.

10.4. Use of Multivariate Adaptive Regression Splines to Model JAR Data

A logical step in analyzing diagnostic data using the JAR scale consists of regressing the JAR scores against liking scores to determine the exact nature of the relationship. However, some difficulties arise from the fact that the relationship between hedonic and JAR scores is not linear. With ordinary least squares, a quadratic term can be added to the model, but the regression coefficients (β coefficients) become difficult to interpret and do not provide a simple answer to the problem at hand, albeit determining the potential effect of non-JAR scores on hedonic scores both above and below the JAR score. The idea behind local nonparametric modeling including MARS (multivariate adaptive regression splines) is to allow for a potentially nonlinear relationship over different ranges or intervals of X (Sephton, 2001).

Friedman (1991) introduced the MARS approach of using smoothing splines to fit the relationship between a vector of predictors (in this case, diagnostic attributes) and a response variable (in this case, a hedonic response). The principle of MARS is to split the predictive variable space into several small distinct intervals (the endpoints of the intervals are called knots), transform the original predictive variables over the interval using basis functions, fit a linear regression model to each interval, and finally, combine individual linear models obtained from each interval as a final MARS model. A general MARS model for a single response and a vector of predictors is described by Equation (10.4):

$$Y = \beta_0 + \beta_1 BF_1(X) + \beta_2 BF_2(X) + \ldots + \beta_M BF_M(X) + \varepsilon, \qquad (\text{Eq. 10.4})$$

where Y is a response variable such as overall liking, X is the vector of predictors (such as the appropriateness of size, color, flavor, salt, thin/thick, stickiness, etc., measured on JAR scales), and BF_k is the kth basis function that represents the transformation of a single predictive variable or multivariable interaction terms over the interval/region, which is similar to a variable combination created by principal components. The basis function is interpreted as the additive and interactive effects of the predictive variables on the knot locations. The variable M is the number of basis functions included in the final model. The regression coefficients β_k are estimated by minimizing the sum of squared residuals ε. Both the predictors to use and the knots for each predictor are found via a brute force, exhaustive search procedure, using very fast update algorithms. Predictors, knots, and interactions are optimized simultaneously by evaluating a loss-of-fit criterion. MARS chooses the loss of fit that most improves the model at each step. In addition to searching variables one by one, MARS also searches for interactions between variables, allowing any degree of interaction to be considered. MARS automates both the selection of variables and the nonparametric transformation of variables to achieve the best model fit. Variable transformation is accomplished implicitly through the piecewise regression function used by MARS to trace arbitrary nonlinear functions. MARS communicates this nonparametric transformation graphically, displaying the predicted response as a function of either one or two variables (Salford Systems Homepage, http://salford-systems.com).

Model selection is based on the generalized cross-validation (GCV) criterion of Craven and Wahba (1979). The GCV can be expressed as follows:

$$GCV = \frac{1}{N} \frac{\sum_{i}^{N}[Y_i - f_M(X_i)]^2}{\left[1 - \frac{C(M)}{N}\right]^2},$$ (Eq. 10.5)

where N is the number of observations;

$$\sum_{i=1}^{N}[Y_i - f_M(X_i)]^2$$

measures the lack of fit on the M basis function model $f_M(X_i)$, which corresponds to the sum of squared residuals; $\{1 - [C(M)/N]\}^2$ is a penalty term for using M basis functions; and $C(M)(\geq M)$ is the cost-complexity measure of a model containing M basis functions.

The optimal MARS model is the one with the lowest GCV value. The selection of the optimal MARS model is performed in forward and backward stepwise fashion. In the forward selection, MARS constructs an overly large model by adding basis functions. As basis functions are added, the model becomes more flexible and more complex, and the process continues until a user-specified maximum number of basis functions is reached. In the backward selection, basis functions are deleted in the order of least contribution to the model until an optimal model is found (MARS, 2001). This is in fact very similar to the well-known concept of stepwise regression analysis.

Special software is needed to carry out the analysis. A free MARS evaluation version can be downloaded at http://www.salford-systems.com/mars.php. Instructions for using the software are given in Appendix 10.2.

Let us now discuss a specific example. In the study at hand, published by Xiong and Meullenet (2004), eight cheese stick appetizers were presented to a consumer group of 160 members from two test locations. A more extensive description of the data is found in Chapter 1. Briefly, each consumer was asked to evaluate the overall acceptance and the acceptance of appearance, flavor, and texture of each sample on a 9-point hedonic scale with 1 = Dislike Extremely and 9 = Like Extremely. Consumers were also asked to rate the color, size, breading, saltiness, crispness, and cheese texture of each sample on a 5-point JAR scale, with the JAR score being 3. The MARS functions for texture acceptance are presented in Table 10.2 and are represented in Figure 10.6. Four basis functions were given by MARS—a pair (BF1 and BF2) for crispiness and another pair (BF3 and BF4) for cheese texture. The regression coefficients of −1.667 for BF1 = max(0, Crispiness − 3) and −1.187 for BF2 = max(0, 3 − Crispiness) mean that the drop rate was faster over the region of 3–5 than over the region of 3–1, indicating that "too crispy" was more

Table 10.2. Basis functions for predicting texture acceptance from diagnostic attributes (observations $n = 1304$).

Number of Knots	Interactions	Basis Functions (Texture Acceptance)	Adj R^2	GCV
4–40	No	BF1 = max(0, Crispiness − 3.0); BF2 = max(0, 3.0 − Crispiness); BF3 = max(0, CheeseTexture − 3.0); BF4 = max(0, 3.0 − CheeseTexture); Y = 7.300 − 1.667 × BF1 − 1.187 × BF2 − 1.155 × BF3 − 1.299 × BF4	0.448	2.310
5	Yes	BF1 = max(0, Crispiness − 3.0); BF2 = max(0, 3.0 − Crispiness); BF3 = max(0, CheeseTexture − 3.0); BF4 = max(0, 3.0 − CheeseTexture); BF6 = max(0, 3.0 − CheeseTexture) × BF2; BF7 = max(0, CheeseTexture − 4.0); BF9 = max(0, Crispiness − 3.0) × BF7; BF11 = max(0, Crispiness − 2.0) × BF7; BF15 = max(0, Crispiness − 1.0) × BF4; Y = 7.328 − 1.756 × BF1 − 1.283 × BF2 − 0.971 × BF3 − 0.576 × BF6 + 2.761 × BF9 − 1.362 × BF11 − 0.681 × BF15	0.459	2.307

Note: Adj R^2 = Adjusted R^2; GCV = generalized cross validation; Y = overall acceptance score.

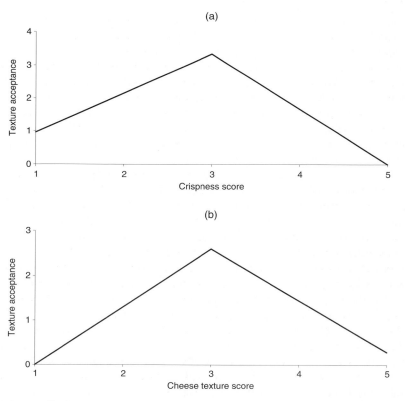

Figure 10.6. Contributions to texture acceptance from crispness (a) and cheese texture (b) JAR scores.

detrimental to the acceptance of texture than "not crispy enough." The regression coefficients of −1.155 for BF3 = max(0, CheeseTexture − 3) and −1.299 for BF4 = max(0, 3 − CheeseTexture) mean that the drop rate was slower over the region of 3–5 than over the region of 3–1, implying that "too soft/melted" had more detrimental effects on the texture acceptance than "too firm/not melted." In terms of MARS relative variable importance values (100% and 96.73%, respectively), crispness made slightly more of a contribution to texture acceptance than did cheese texture. The R^2 between the observed and predicted mean scores of texture acceptance was 0.95, indicating that the fitted MARS model could accurately predict the mean scores for texture acceptance. The use of interactions in the model did not improve the R^2 and GCV values, but the models became more complicated. Therefore, the discussion of the interaction model is not given here.

The advantages of the novel modeling technique MARS is that it automates both the selection of variable and the nonparametric transformation of variables (basis functions) to achieve the best-fit model, and variable transformation is accomplished implicitly through the piecewise linear regression function to express underlying nonlinear functions. MARS, therefore, has potential value in the modeling of consumer acceptance of food, especially in exploring and finding the complex relationships between response and predictors and in relating predictors evaluated on a JAR scale to a response such as consumer acceptance.

10.5. A Proportional Odds/Hazards Approach to Diagnostic Data Analysis

At present there are three popular methods for testing for significant differences among distributions of JAR scores (i.e., the Stuart Maxwell/McNemar test, The Chi-square [χ^2]/McNemar test, and the t-tests/ANOVA). The Stuart Maxwell/McNemar test is a two-stage test procedure that is unable to test the overall effect of products on a JAR response. Both the Stuart Maxwell and the McNemar tests require the original JAR response categories to be collapsed into three and two categories, respectively, which results in loss of some useful information present in the original categories. The JAR categories are ordinal in nature, but the Stuart Maxwell and the χ^2 tests do not take this ordinal nature into account, resulting in a loss of power for ordinal data. In addition, the McNemar test was designed for matched pairs of samples and not for independent samples. Furthermore, the McNemar test is not appropriate for unbalanced data (e.g., the numbers of observations are different for different products). For diagnostic purposes, 3-, 5-, or even 7-point JAR scales are often used, and the data from these scales should be treated as categorical or ordinal. Ordinal or categorical data are not normally distributed. The t-test/ANOVA requires the data to be on either an interval or a ratio scale and may not be appropriate for seven or fewer JAR response categories because the normal assumption may fail.

This brings us to the real subject of this section. An innovative alternative method to those tests is the proportional odds/hazards model. The proportional odds and hazards models are widely used in medicine and life science and survey studies. Recently, both models have been applied to the sensory field for preference mapping (Meullenet et al. 2003) and shelf-life studies (Gimenez et al. 2003), respectively. In this section, we first present details of proportional odds and hazard models and then apply the concepts to a specific example. The sample data set we use for the example was provided by the Committee E18 on Sensory Evaluation of Materials and Products (subcommittee E18.04) for the development of a series of case studies on the analysis of JAR data.

10.5.1. Proportional Odds Model

The proportional odds model (POM) is widely used to analyze the structure (frequencies) of ordered categorical data and can be mathematically expressed as (Agresti, 1990):

$$\text{logit}(\pi_{ijk}) = \ln\left(\frac{\pi_{ijk}}{1-\pi_{ijk}}\right) = \alpha_k + \beta' X_{ij}, \quad \text{(Eq. 10.6)}$$

where $\ln[\pi_{ijk}/(1-\pi_{ijk})]$ is the log of the odds of response Y_{ij} (scores) being in category k or below ($k = 1, 2, \ldots, K-1$, with $K = 5$ for a 5-point JAR scale, also referred to as the logit). The equation $\pi_{ijk} = P(Y_{ij} \leq k \mid X_{ij})$ is the probability the categorical score (response Y_{ij}) rated by the ith panelist ($i = 1, 2, \ldots, n$, n is the total number of panelists) will be in category k or below for the sample j, given a vector of independent variables X_{ij}. The variable α_k is an intercept parameter representing the logits of the cumulative response when the explanatory variables X_{ij} are zero and $\alpha_1 \leq \alpha_2 \leq \ldots \leq \alpha_{K-1}$. The term β is a vector of unknown parameters representing the effect of X_{ij} on the response. POM is implemented conveniently by the LOGISTIC procedure in SAS with LINK = LOGIT, where the equal slopes assumption (also called parallelism assumption), goodness of fit, and overall effect of treatments/products can be tested. POM is appropriate only if the parallelism assumption is satisfied. Paired comparison tests between treatments/products can be performed using the CONTRAST statements in the procedure. The SAS codes for POM are given in Appendix 10.3.

10.5.2. Proportional Hazards Model

The Cox proportional hazards models (PHM) are widely used in survival analysis and can be expressed as

$$S(k, X_{ij}) = S_0(k) e^{-e^{\beta X_{ij}}}, \quad \text{(Eq. 10.7)}$$

where $S_0(k) = e^{-e^{\beta X_{ij}}}$ is the baseline survivor function and $S(k, X_{ij}) = 1 - \pi_{ijk}$ is the survivor function. The model is reexpressed as:

$$1 - \pi_{ijk} = e^{-e^{\alpha_k}} e^{-e^{\beta X_{ij}}} = e^{-e^{\alpha_k + \beta X_{ij}}} \quad \text{(Eq. 10.8)}$$

By taking ln twice, we have

$$\ln[-\ln(1 - \pi_{ijk})] = \alpha_k + \beta' X_{ij} \quad \text{(Eq. 10.9)}$$

which becomes a linear regression model using the log-log link function LINK = CLOGLOG. The SAS program for fitting this model is provided in Appendix 10.4.

10.5.3. Comparison of POMs and Proportional Hazard Models

The POMs and proportional hazards models (PHMs) have the same underlying assumption, but they use different link functions to model ordinal response data. The comparison of POM and PHM is presented in Table 10.3. The goodness of fit for both POM and PHM is assessed by the likelihood ratio or deviance G^2.

Table 10.3. Comparison of the proportional odds and hazards models.

	Proportional Odds Model	**Proportional Hazards Model**
Assumption	Equal slopes across levels of a response variable	Equal slopes across levels of a response variable
Model	$P[Y \leq k] = \dfrac{1}{1+e^{-(\alpha_k + \beta' x)}}$	$P[Y \leq k] = 1 - e^{-e^{\alpha_k + \beta' x}}$
Link function	Logit	Complementary log-log

10.5.4. Examples

The POMs and PHMs were fitted to the cumulative frequencies of each JAR attribute using the SAS LOGISTIC procedure, and the results are separately discussed in the following.

For the JAR attribute size, the chi-square (χ^2) for testing the equal slopes assumption was 7.623 with a p value of 0.814, which was not significant with respect to a chi-square distribution with 12 degrees of freedom (df) at a significance level (α) of 0.05. This indicated that the parallelism assumption was satisfied. The likelihood ratio (deviance) G^2 was 7.132 (df = 12) with $p = 0.849$, indicating that the POM adequately fitted the data. The parameterization used in the SAS system is one that leaves out the parameter for the baseline (sample 170 in this case) with which each sample is compared. Hence, a positive parameter estimate (β) in Table 10.4 means that sample 170 was larger in size than the compared sample, whereas a negative estimate means that sample 170 was smaller in size. The p value for the chi-square statistic is used to test whether the difference between the compared sample and the baseline sample is significant. Because all the p values were greater than $\alpha = 0.05$ (Table 10.4), all the samples were not significantly different in size from sample 170. Overall, the effect of products was not significant ($\chi^2 = 3.273$, df = 4, p value = 0.513) at $\alpha = 0.05$. This indicated that all the products had similar distributions of their size scores evaluated on a 5-point JAR scale. An advantage of the POM over the Stuart Maxwell test is that, as for ANOVA, an overall significance test among all samples can be performed. However, ANOVA requires the JAR response to be continuous, whereas POM does not because it models the cumulative frequency of the JAR responses. The χ^2 test can be used to test the overall effect of products but ignores the ordinarity of the JAR response.

For the JAR attribute color, the parallelism (equal slopes) assumption was not met for POM ($\chi^2 = 15.900$, df = 8, p value = 0.044) at $\alpha = 0.05$ but was met at $\alpha = 0.01$. If $\alpha = 0.01$ were used, the overall effect of products was then significant ($\chi^2 = 19.944$, df = 4, p value = 0.0005) at the significance level of 0.05. This indicated that not all the products had similar distributions of color JAR scores. This means that products that were significantly different in color need to be identified. By contrasting a pair of products (e.g., see the sample SAS codes in Appendix 10.3 on how to syntax this), like a t-test, the difference between two products can be determined using the Wald chi-square in the SAS LOGISTIC procedure. For more than two products, this paired comparison test is performed repeatedly for all possible pairs of products. The number of paired comparison tests will rapidly increase as the number of products increases. The major problem with the paired comparison tests is that the chance of Type I error (rejecting H_0 when it is true) will go up as the number of the paired comparison tests increases. The overall level of significance for a set of paired comparison tests is much higher than that for each single test. To maintain the prespecified overall level of significance, each test must work at a lower

Table 10.4. Parameter estimates from the proportional odds model for the JAR attribute size.

Parameter	Estimate	Standard Error	Chi-Square	p value
Intercept1	−3.048	0.2597	137.769	<0.0001
Intercept2	−1.181	0.1916	37.990	<0.0001
Intercept3	0.359	0.1843	3.799	0.0513
Intercept4	2.564	0.2397	114.426	<0.0001
Sample 458	0.382	0.2544	2.255	0.1332
Sample 523	0.081	0.2543	0.102	0.7495
Sample 896	−0.026	0.2544	0.010	0.9197
Sample 914	0.138	0.2542	0.296	0.5866

Table 10.5. Estimates of parameters and odds ratios from the proportional odds model for color evaluated on a 5-point JAR scale.

Effect	Estimate	Standard Error	Odds Ratio	Wald Chi-square	p value
Intercept1	−6.629	1.0450		40.246	<0.0001
Intercept2	−2.431	0.3266		55.393	<0.0001
Intercept3	3.458	0.3926		77.568	<0.0001
Sample 458 versus 170	0.575	0.4208	1.777	1.867	0.1718
Sample 523 versus 170	0.091	0.4408	1.095	0.043	0.8363
Sample 896 versus 170	−0.713	0.4653	0.490	2.347	0.1255
Sample 914 versus 170	1.124	0.4008	3.078	7.869	0.0050
Sample 523 versus 458	−0.484	0.4156	0.6164	1.355	0.2444
Sample 896 versus 458	−1.288	0.4547	0.276	8.020	0.0046
Sample 914 versus 458	0.549	0.3639	1.732	2.279	0.1311
Sample 896 versus 523	−0.804	0.4642	0.448	2.999	0.0833
Sample 914 versus 523	1.033	0.3947	2.810	6.850	0.0089
Sample 914 versus 896	1.837	0.4415	6.278	17.311	<0.0001

level of significance. This is usually achieved by dividing the overall significance level by the total number of possible comparisons, namely, $\alpha' = \alpha/m$, where α' is the significance level for each single paired comparison test, α is the prespecified overall significance level (usually $\alpha = 0.05$), and m is the total number of possible paired comparison tests, which is calculated as $n(n − 1)/2$ (n is the number of products or treatments). For the five products in this case study, the total number of possible comparisons is $5(5 − 1)/2 = 10$, when $\alpha = 0.05$, $\alpha' = 0.05/10 = 0.005$, which was used as the significance level for each following single paired comparison test.

The parameter estimates and odds ratios between all pairs of the samples are then obtained from contrasts and are presented in Table 10.5. The p values in Table 10.5 are used to test if a pair of samples was significantly different at $\alpha' = 0.005$. For example, the p value for the pair of samples 458 and 170 is $0.1718 > \alpha' = 0.005$, indicating that the two samples are not significantly different in color (i.e., meaning the JAR score distributions are not significantly different from each other, not that the products were identical in color). The p values for the pairs of samples 896 versus 458 and 914 versus 896 are 0.0046 and <0.0001, respectively, which indicates that sample 896 has a distribution of JAR color scores significantly different from those of samples 458 and 914. As mentioned above, the signs of parameter estimates can be used to determine the directional difference between two products. Sample 896 is overall significantly darker in color than sample 458 because of the negative parameter estimate of −1.288, and sample 914 is significantly lighter in color than sample 896 because of the positive estimate of 1.837. As a result, sample 896 is significantly darker in color than samples 458 and 914, and other pairs of samples are not significantly different from each other. The contrasting method provided for the POM in the SAS LOGISTIC procedure is another advantage over the two-stage test procedure of the chi-squared/McNemar tests. The interpretation of parameters is usually done using odds ratios. For example, the odds ratio of 6.278 (= $e^{1.837}$, with 1.873 being the parameter estimate) for samples 914 versus 896 (Table 10.5) means that the odds of consumer rating sample 914 as "too light" in color is 6.278 times the odds for sample 896, so consumers rated sample 914 lighter in color than sample 896.

If $\alpha = 0.05$ had been chosen for the significance level, the equal slopes assumption for POM would not be met because the p value of 0.044 was less than 0.05. This indicates that the proportional odds model could fail in some cases if the parallelism assumption is not satisfied. If POM fails, the PHM could be used as an alternative. For the JAR attribute color, the parallelism assumption is met for PHM ($\chi^2 = 14.014$, df = 8, p value = 0.081) at $\alpha = 0.05$. The overall effect of products is significant ($\chi^2 = 16.875$, df = 4, p value = 0.002) at the significance level of 0.05, indicating that some products have significantly different distributions of JAR color scores. The parameter estimates for PHM are provided in Table 10.6. As for POM, a positive parameter estimate for PHM (Table 10.6) means that sample 170 is "darker" in color than the compared sample, whereas a negative estimate means that sample 170 is "lighter" in color. The p values show that samples 914 versus 170, 914 versus 523, and 914 versus 896 are significantly different from each other ($\alpha' = 0.005$). Sample 914 exhibited lower color JAR scores (i.e., it was lighter) than samples 170, 523, and 896. A comparison of Tables 10.5 and 10.6 shows that results from both POM and PHM are different. However, the significant contrasts observed for POMs are also found to be significant for PHMs. In this case, we would tend to trust the

Table 10.6. Parameter estimates from the proportional hazards model for color assessed on a 5-point JAR scale.

Effect	Estimate	Standard Error	Wald Chi-Square	p value
Intercept1	−6.654	1.0138	43.078	<0.0001
Intercept2	−2.539	0.2106	145.381	<0.0001
Intercept3	1.099	0.1384	62.957	<0.0001
Sample 458 versus 170	0.477	0.2323	4.213	0.0401
Sample 523 versus 170	0.059	0.1960	0.092	0.7618
Sample 896 versus 170	0.011	0.1937	0.003	0.9569
Sample 914 versus 170	1.076	0.2943	13.377	0.0003
Sample 523 versus 458	−0.417	0.2339	3.185	0.0743
Sample 896 versus 458	−0.466	0.2326	4.021	0.0449
Sample 914 versus 458	0.599	0.2979	4.050	0.0442
Sample 896 versus 523	−0.049	0.1964	0.062	0.8032
Sample 914 versus 523	1.017	0.2944	11.928	0.0006
Sample 914 versus 896	1.066	0.294	13.115	0.0003

Table 10.7. Estimates of parameters and odds ratios from the proportional odds model for flavor.

Effect	Estimate	Standard Error	Odds Ratio	Wald Chi-Square	p value
Intercept1	−5.043	0.4356		134.040	<0.0001
Intercept2	−2.167	0.2517		74.128	<0.0001
Intercept3	1.579	0.2314		46.586	<0.0001
Intercept4	3.522	0.3207		120.594	<0.0001
Sample 458 versus 170	−0.855	0.3010	0.425	8.072	0.0045
Sample 523 versus 170	−0.397	0.3057	0.672	1.685	0.1942
Sample 896 versus 170	−0.018	0.3096	0.982	0.003	0.9534
Sample 914 versus 170	2.268	0.3153	9.657	51.718	<0.0001
Sample 523 versus 458	0.458	0.2919	1.582	2.466	0.1163
Sample 896 versus 458	0.837	0.3006	2.310	7.756	0.0054
Sample 914 versus 458	3.123	0.3253	22.713	92.177	<0.0001
Sample 896 versus 523	0.379	0.3054	1.461	1.538	0.2149
Sample 914 versus 523	2.665	0.3218	14.361	68.560	<0.0001
Sample 914 versus 896	2.286	0.3157	9.833	52.421	<0.0001

results from PHMs because the parallelism assumption was met at $\alpha = 0.05$ for PHM but not for POM. A disadvantage of PHM is that it does not provide odds ratios for the interpretation of parameters.

For the JAR attribute flavor, the parallelism assumption is satisfied for POM ($\chi^2 = 20.425$, df = 12, p value = 0.06) at $\alpha = 0.05$. The overall effect of products is significant ($\chi^2 = 105.198$, df = 4, p value < 0.0001), indicating that the products are not from the same population for JAR flavor. The p values (Table 10.7) show that sample 914 has significantly lower JAR flavor scores than all other samples because of the positive estimates, whereas sample 458 has significantly higher JAR flavor scores than samples 170, 896, and 914.

Similar to ANOVA, the proportional odds/hazards model can be used to test overall differences among products and directional differences between samples but does not require the JAR responses to be normally distributed, unlike ANOVA. For a 5-point JAR scale data, the normal assumption may fail for ANOVA. The proposed method with the SAS LOGISTIC procedure is easy to use and implement and provides a viable alternative to analyzing JAR data. However, the proportional odds and hazards models are valid only if the equal slopes assumption is met. In some cases, both POM and PHM will fail, especially if a large number of products are involved in the analysis.

10.6. Use of Dummy Variables to Model JAR Data

Penalty analysis is a graphical technique that reveals the possible penalty paid by a product in terms of reduced overall liking by not being at a "just right" level for a specific modality. This penalty is often called mean drop on overall liking. Because this type of analysis provides the sensory scientist with a prioritized list of critical product characteristics that most penalize product performance, penalty analysis is widely used in consumer testing and market research. However, it is not a regression-based method, ignores correlations among product characteristics, and cannot be used to predict consumer overall acceptance from JAR data. In addition, the mean drop estimated by penalty analysis for a specific attribute is not the estimate of the "true" mean drop on overall liking. MARS is a regression method and can be used to estimate the "true" mean drop, but it requires the assumption of independence of the JAR variables. Because JAR variables are often correlated with each other, this limits the use of MARS. A dummy variable approach with two models (i.e., analysis of covariance and partial least squares regression) is proposed as an alternative to penalty analysis and MARS.

10.6.1. Analysis of Covariance with Dummy Variables

Analysis of covariance is a combination of ANOVA and regression analysis. It is used to assess the statistical significance of mean differences among treatment groups with an adjustment made for initial differences on one or more covariates. When analysis of covariance applies to the prediction of overall liking from JAR variables, the covariance analysis model is expressed as

$$Y_{ij} = \mu_i + \sum_{k=1}^{p} \beta_{ik} X_{ijk} + \eta_{ij} + \varepsilon_{ij} \quad (i = 1, 2 \ldots, t; j = 1, 2 \ldots, m), \quad \text{(Eq. 10.10)}$$

where the response variable Y_{ij} is the overall liking score given by the jth consumer for the ith product; μ_i is the mean of the response variable for the ith product; independent variable X_{ijk} is the JAR score given by the jth consumer for the kth JAR variable of the ith product; t, m, and p are the numbers of products, consumers, and JAR variables used in the test, respectively; β_{ik} is the regression coefficient for the kth JAR variable and the ith product; and η_{ij} and ε_{ij} are the random effect term and residual for the jth consumer and the ith product, respectively. This model (Equation [10.10]) is unable to correctly describe the nonlinear relationship between the response variable (Y_{ij}) and the covariates ($X_{ij1}, X_{ij2}, \ldots, X_{ijp}$) because the JAR scale has its "best/ideal" score in the middle (Figure 10.7a) of the scale. Take a 5-point JAR scale (1–5) as an example. The best/ideal score, also called the JAR score, is 3. As the scores of a JAR variable (X_{ijk}) are far away from the JAR score over both sides/regions (both Too Little and Too Much), consumer acceptance scores (Y_{ij}) would be expected to drop or stay constant (Figure 10.6). It is possible that the drop rates over both regions of the JAR score may be different. To describe this phenomenon, two dummy variables (Z_{ijk1} and Z_{ijk2}) are introduced to represent each original JAR variable (X_{ijk}). The presentation scheme is provided in Table 10.8.

Over the region (1–3) of X_{ijk}, Z_{ijk1} changes from –2 to 0, and Z_{ijk1} is 0 over the region (3–5). In contrast, Z_{ijk2} is 0 over the region (1–3) and changes from 0 to 2 over the region (3–5). If the drop rates over the two regions are the same, Z_{ijk1} and Z_{ijk2} are combined into a single dummy variable $Z_{ijk} = Z_{ijk1} - Z_{ijk2}$ to represent X_{ijk}.

The replacement of the original JAR variables in Equation (10.10) by the dummy variables yields the following model:

$$Y_{ij} = \mu_i + \sum_{k=1}^{p} (\alpha_{ik1} Z_{ijk1} + \alpha_{ik2} Z_{ijk2}) + \eta_{ij} + \varepsilon_{ij} \quad (i = 1, 2 \ldots, t; j = 1, 2 \ldots, m), \quad \text{(Eq. 10.11)}$$

where the pair (Z_{ijk1}, Z_{ijk2}) are dummy variables for the original JAR variable X_{ijk} and α_{ik1} and α_{ik2} are regression coefficients for the kth pair of dummy variables, respectively. For each pair of regression coefficients (α_{ik1}, α_{ik2}), the sign (+) of α_{ik1} must be opposite to the sign (–) of α_{ik2}, as shown in Figure 10.7b. Results yielding α_{ik1} and α_{ik2} of the same sign would indicate that the original JAR variables ($X_{ij1}, X_{ij2}, \ldots, X_{ijp}$) are highly correlated with one another or that there is noise in the data (e.g., an inconsistency of scores among judges). If the results yield variables with the same sign and it is determined to be a result of noise in the data, the dummy variables with unexpected signs should be removed from

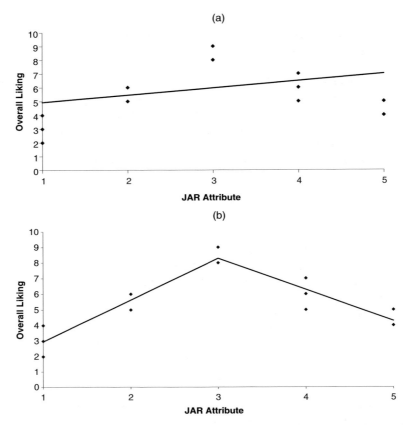

Figure 10.7. Examples of linear regression models using the original variable (a) and dummy variables (b) to map relationships between JAR and hedonic scores.

Table 10.8. Scheme for using two dummy variables (Z_1 and Z_2) or one dummy variable (Z) to represent one JAR variable (X) on a 5-point or 7-point JAR scale.

5-Point JAR Scale				7-Point JAR Scale			
X	Z_1	Z_2	Z	X	Z_1	Z_2	Z
1	−2	0	−2	1	−3	0	−3
2	−1	0	−1	2	−2	0	−2
3	0	0	0	3	−1	0	−1
4	0	1	−1	4	0	0	0
5	0	2	−2	5	0	1	−1
				6	0	2	−2
				7	0	3	−3

the model (Equation [10.11]). If this is a result of correlations between the variables, the model (Equation [10.11]) is not appropriate. As such, this covariance analysis model (Equation [10.11]) is appropriate for establishing relationships between JAR and liking variables only if the original JAR variables ($X_{ij1}, X_{ij2}, \ldots, X_{ijp}$) are independent of each other. Correlation coefficients can be used to check the independence between JAR variables. A stepwise method (such as backward elimination) can be applied to this model for selecting important variables. If Equation (10.11) is appropriate, the term ($\alpha_{ij1} Z_{ijk1} + \alpha_{ij2} Z_{ijk2}$) is always either zero or negative. If the mean of the terms ($\alpha_{ij1} Z_{ijk1} + \alpha_{ij2} Z_{ijk2}$) ($j = 1, 2, \ldots, m$) across all consumers is zero, it indicates that there is no significant relationship between the kth JAR variable and the response (e.g., overall liking). By contrast, a negative mean can be explained as the estimate of the mean drop on overall liking resulting from the kth JAR variable not being JAR for the ith product. Because Equation (10.11) includes all dummy variables involved, it is usually called the full model. On the basis of the full model, many hypotheses can be formed and tested using an F test. For example, if it is assumed that the effect of each dummy variable on the response is the same (common slope) for all products, then the model (Equation [10.12]) becomes

$$Y_{ij} = \mu_i + \sum_{k=1}^{p}(\alpha_{k1}Z_{ijk1} + \alpha_{k2}Z_{ijk2}) + \eta_{ij} + \varepsilon_{ij} \quad (i = 1, 2 \ldots, t; j = 1, 2 \ldots, m). \quad \text{(Eq. 10.12)}$$

This model (Equation [10.12]) is called the reduced model because it contains only a subset of the variables used in the full model. An F test is used to determine which model (the full model or the reduced model) fits the data best. The testing hypotheses are given here: H_0: reduced model (Equation [10.12]) = full model (Equation [10.11]); H_a: full model > reduced model.

The H_0 would be rejected if the calculated $F > F_{\alpha, t-1, N-t}$, where the F value is calculated by

$$F = \frac{(SSE_R - SSE_F)/(DF_R - DF_F)}{SSE_F/DF_F}, \quad \text{(Eq. 10.13)}$$

where SSE_R and SSE_F are the sum of squares of errors for the reduced and full models, respectively, and DF_R and DF_F are the degrees of freedom of the error term for the reduced and full models, respectively. Similarly, F tests can be used to test whether a pair or some pairs of dummy variables have the same effect on the response for all products.

During the analysis of attributes assessed of a JAR scale, one is often concerned with assessing the drop rates over two regions of the scale (Too Little and Too Much) and determining whether these drop rates are significantly different from each other. The following hypothesis can then be tested:

H_0: $\alpha_{ik1} = -\alpha_{ik2}$, or $|\alpha_{ik1}| = |\alpha_{ik2}|$ for all pairs of dummy variables;
H_a: $\alpha_{ik1} \neq -\alpha_{ik2}$, or $|\alpha_{ik1}| \neq |\alpha_{ik2}|$ for at least some pairs of dummy variables. (Eq. 10.14)

Under the null hypothesis (H_0), the model (Equation [10.12]) can be simplified as

$$Y_{ij} = \mu_i + \sum_{k=1}^{p}\alpha_{ik1}(Z_{ijk1} - Z_{ijk2}) + \eta_{ij} + \varepsilon_{ij} = \mu_i + \sum_{k=1}^{p}\alpha_{ik1}Z_{ijk} + \eta_{ij} + \varepsilon_{ij} \quad \text{(Eq. 10.15)}$$
$$(i = 1, 2 \ldots, t; j = 1, 2 \ldots, m),$$

where $Z_{ijk} = Z_{ijk1} - Z_{ijk2}$ (see Table 10.8). By using the above F test (Equation [10.15]), the full (Equation [10.11]) and reduced (Equation [10.15]) models can be compared to determine whether two drop rates over the two JAR regions are the same for all pairs of dummy variables simultaneously. Similarly, an F test can be used to test whether the drop rates are the same for a pair or some pairs of dummy variables or whether all products have the same means. If other factors (such as gender, age, etc.) are of interest, they can be added to the models. Once the final model is determined, the mean overall liking score (i.e., any other hedonic attribute can be used) for each individual product can be predicted by

$$\overline{Y}_i = \mu_i + \sum_{k=1}^{p}(\alpha_{ik1}\overline{Z}_{ik1} + \alpha_{ik2}\overline{Z}_{ik2}) \quad (i = 1, 2\ldots, t) \quad \text{(Eq. 10.16)}$$

where $\overline{Y}_i$ is the predicted mean score of overall liking for the ith product; $\overline{Z}_{ik1}$ and $\overline{Z}_{ik2}$ are the means of dummy variables Z_{ijk1} and Z_{ijk2} cross consumers for the ith product and the kth JAR variable, respectively; $\alpha_{ik1}\overline{Z}_{ik1}$ and $\alpha_{ik2}\overline{Z}_{ik2}$ are defined as the estimates of the mean drop on overall liking resulting from the kth JAR variable being either Too Little or being Too Much, respectively. The estimates of the mean drop on overall liking for each JAR variable are often useful in product development. As was pointed out previously,

$$\sum_{k=1}^{p}(\alpha_{ik1}\overline{Z}_{ik1} + \alpha_{ik2}\overline{Z}_{ik2}) \leq 0 \quad \text{(Eq. 10.17)}$$

is always true if the model fits the data appropriately. For the ith product,

$$\sum_{k=1}^{p}(\alpha_{ik1}\overline{Z}_{ik1} + \alpha_{ik2}\overline{Z}_{ik2}) = 0 \quad \text{(Eq. 10.18)}$$

holds only if $\overline{Z}_{ik1} = \overline{Z}_{ik2} = 0$ for all k ($k = 1, 2, \ldots, q$), which represents that all attributes of the ith product are JAR. From the above discussion, the covariance analysis model using dummy variables is very flexible for identifying attributes (i.e., from JAR scales) penalizing liking.

10.6.2. Partial Least Squares Regression with Dummy Variables

In sensory evaluation, it is often found that some variables are highly correlated with each other. This correlation or dependence between variables violates the independence assumption for covariance analysis models, so analysis of covariance is no longer valid. Partial least squares (PLS) regression or principal component regression (PCR) are often used to handle this type of collinearity problems. For a single response, PLS regression models for each individual product can be expressed as

$$Y_j = \beta_0 + \beta_1 PC_{j1} + \beta_2 PC_{j2} + \ldots + \beta_s PC_{js} + \varepsilon_j \quad (j = 1, 2, \ldots, m; s \leq p) \quad \text{(Eq. 10.19)}$$

where the response variable Y_j can be the overall liking score given by the jth consumer to a product, PC_{jk} is the score of the kth principal component for the jth consumer, s is the number of principal components and is less than or equal to p number of the original variables, β_0 is the regression intercept, β_k is regression coefficient for the kth PC, and β_j is the residual for the jth consumer. If the original JAR variables are used to identify the drivers of liking for each product (not all products), principal components (PCs) are calculated as follows

$$PC_{jk} = a_{k1}X_{j1} + a_{k2}X_{j2} + \ldots + a_{kp}X_{jp} \quad (j = 1, 2, \ldots, m; k = 1, 2, \ldots, s) \quad \text{(Eq. 10.20)}$$

where a_{kl} ($k = 1, 2, \ldots, s; l = 1, 2, \ldots, p$) is the loading for the kth PC and the lth original JAR variable (X_l), and X_{jl} is the score given by the jth consumer for the lth JAR variable. As discussed previously, the nonlinear relationships between the original JAR variables and the response (e.g., overall liking) cannot be appropriately described by a linear regression model (such as PLS or PCR) using the original variables. Dummy variables can be used as previously described in PLS or PCR models to estimate these nonlinear relationships, and PCs using dummy variables are calculated as

$$PC_{jk} = (b_{k11}Z_{j11} + b_{k12}Z_{j12}) + (b_{k21}Z_{j21} + b_{k22}Z_{j22}) + \ldots + (b_{kp1}Z_{jp1} + b_{kp2}Z_{jp2}) \quad \text{(Eq. 10.21)}$$
$$(j = 1, 2, \ldots, m; k = 1, 2, \ldots, s)$$

where the pair (b_{kl1}, b_{kl2}) $(k = 1, 2, \ldots, s; l = 1, 2, \ldots, p)$ are the loadings for the kth PC and the lth pair of dummy variables (Z_{l1}, Z_{l2}), which represents the lth JAR variable (X_l). By substituting PC_{jk} in Equation (10.21) into the PLS model [Equation (10.19)], we obtain the following PLS model using p pairs of dummy variables.

$$Y_j = \gamma_0 + (\gamma_{11}Z_{j11} + \gamma_{12}Z_{j12}) + (\gamma_{21}Z_{j21} + \gamma_{22}Z_{j22}) + \ldots + (\gamma_{p1}Z_{jp1} + \gamma_{p2}Z_{jp2}) + \varepsilon_j \quad \text{(Eq. 10.22)}$$
$$(j = 1, 2, \ldots, m)$$

where γ_0 is the regression intercept, which can be interpreted as the estimated mean of overall liking for the product if all JAR attributes are JAR, and the pair $(\gamma_{l1}, \gamma_{l2})$ are regression coefficients for the lth pair of dummy variables (Z_{l1}, Z_{l2}), respectively. If the scheme in Table 10.8 is used for each pair of dummy variables, regression coefficient γ_{l1} should be positive, and γ_{l2} should be negative, which means that the response (e.g., overall liking) has its maximum at the JAR score. If the signs for γ_{l1} and γ_{l2} are not expected ones because of data noise, then the corresponding dummy variables need to be removed from the PLS model as they would probably not be interpretable. Because not all variables in Equation (10.21) have equal influences on the response, those unimportant or insignificant variables to the response need to be removed from the PLS model.

The jackknife optimization method is one of popular methods used to remove unimportant variables from PLS models or to select important variables for PLS models. Selection of important dummy variables for the PLS model by the jackknife method can be done with the Unscramber software program (Unscrambler, version 9.0, CAMO, Norway). The PLS models using q ($q \leq p$) important pairs of dummy variables are given by

$$Y_j = \gamma'_0 + (\gamma'_{j11}Z'_{j11} + \gamma'_{j12}Z'_{j12}) + (\gamma'_{j21}Z'_{j21} + \gamma'_{j22}Z'_{j22}) + \ldots + \quad \text{(Eq. 10.23)}$$
$$(\gamma'_{q1}Z'_{jq1} + \gamma'_{q2}Z'_{jq2}) + \varepsilon_j \quad (j = 1, 2, \ldots, m)$$

where γ'_0 is the regression intercept, and the pair $(\gamma'_{l1}, \gamma'_{l2})$ are regression coefficients for the lth important pair of dummy variables (Z'_{l1}, Z'_{l2}). A pair of dummy variables is defined to be important or significant if at least one of the paired dummy variables is selected into the PLS model by the jackknife or other optimization methods. Important pairs of dummy variables imply that the corresponding JAR attributes are not JAR, whereas unimportant pairs of dummy variables can be interpreted as not having a significant effect on overall liking (or other response) or as being at a JAR level. When the regression coefficients $(\gamma'_{l1}, \gamma'_{l2})$ are not equal for a pair of dummy variables, the JAR attribute is more detrimental to the response (i.e., usually overall liking) over one JAR region than the other. The PLS model [Equation (10.23)] is called the F-model (or pseudo full model) because it includes all important pairs of dummy variables. Based on this F-model, various null hypotheses can be formed and tested just like for the covariance analysis models. For example, if the null hypothesis is that the two drop rates (in absolute values) over the two JAR regions are the same for all pairs of dummy variables in Equation (10.23) [i.e., $H_0: \gamma'_{l1} = -\gamma'_{l2} = \phi'_l$ for all important pairs, $l = 1, 2, \ldots, q$; see Equation (10.15)], then a pair of two dummy variables can be combined into one single dummy variable ($Z'_l = Z'_{l1} - Z'_{l2}$), and Equation (10.11) becomes

$$Y_j = \phi'_0 + \phi'_1 Z'_{j1} + \phi'_2 Z'_{j2} + \ldots + \phi'_q Z'_{jq} + \varepsilon_j \quad (j = 1, 2, \ldots, m) \quad \text{(Eq. 10.24)}$$

where $\phi'_0, \phi'_1, \phi'_2, \ldots, \phi'_q$ are regression coefficients for Equation (10.12). This PLS model [Equation (10.24)] is called the R-model (pseudo reduced model) because it contains only a subset of the all dummy variables used in the above F-model [Equation (10.23)]. Similarly, if the null hypothesis of interest is that the two drop rates are the same only for the first pair of dummy variables (Z'_{j1}, Z'_{j2}), then the PLS model is given by

$$Y_j = \phi'_0 + \phi'_1 Z'_{j1} + (\phi'_{21} Z'_{j21} + \phi'_{22} Z'_{j22}) + \ldots + (\phi'_{q1} Z'_{jq1} + \phi'_{q2} Z'_{jq2}) + \varepsilon_j \quad \text{(Eq. 10.25)}$$
$$(j = 1, 2, \ldots, m)$$

where $\phi'_0, \phi'_1, \phi'_2, \ldots, \phi'_q$ are regression coefficients for Equation (10.13). The PLS model [Equation (10.25)] is another R-model. Unlike the above analysis of covariance, however, there is no F-test available for testing the null hypothesis for comparing the F-model and the R-model. The root mean square error (RMSE) statistic could be used to assess which model is more appropriate. If RMSE values are "substantially" different between the F- and R-models, then the F-model is more appropriate. Otherwise, the R-model is more appropriate on the basis of parsimony. Because the F- and R-models fit the same data, the residuals from the two models are correlated with each other. In addition, RMSE$_R$ (RMSE for R-model) is at least equal to or greater than RMSE$_F$ (RMSE for F-model) because the R-model uses fewer dummy variables than F-model. A paired t-test described by Snedecor and Cochran (1967) for comparing two correlated variances can be used to test whether RMSE$_R$ and RMSE$_F$ are significantly different at a significance level of $\alpha = 0.05$. The t value for the paired t-test is computed as

$$t = r_{FR} \sqrt{\frac{N-2}{1-r_{FR}^2}}; \quad \text{(Eq. 10.26)}$$

where

$$r_{FR} = \frac{F-1}{\sqrt{(F+1)^2 - 4r^2 F}}; \quad \text{(Eq. 10.27)}$$

$$F = \frac{s_R^2}{s_F^2} = \frac{DF_R}{DF_F} \left(\frac{RMSE_R}{RMSE_F}\right)^2; \quad \text{(Eq. 10.28)}$$

N is the number of observations; r is the correlation coefficient between the residuals for F- and R-models; S_R^2 and S_F^2 are standard deviations of the residuals for the R- and F-models, respectively; and DF_R and DF_F are the degrees of freedom for the R- and F-models, respectively, and they are calculated as the difference between the number of observations and the number of principal components in the PLS model. If the computed t value is equal to or greater than the table t value (one-tailed test, degree of freedom $df = N - 2$) at $\alpha = 0.05$, RMSE$_R$ is significantly larger than RMSE$_F$, indicating that the

F-model is more appropriate. If the computed t value is less than the table t value at $\alpha = 0.05$, RMSE_R is not significantly larger than RMSE_F, indicating that the R-model is appropriate. Once the final PLS model is determined, the mean score of overall liking for the product can be predicted by

$$\overline{Y}' = \gamma'_0 + (\gamma'_{11}\overline{Z}'_{11} + \gamma'_{12}\overline{Z}'_{12}) + (\gamma'_{21}\overline{Z}'_{21} + \gamma'_{22}\overline{Z}'_{22}) + \ldots + (\gamma'_{q1}\overline{Z}'_{q1} + \gamma'_{q2}\overline{Z}'_{q2}) \quad \text{(Eq. 10.29)}$$

where $\overline{Y}'$ is the predicted mean score of the response (i.e., usually overall liking); $\overline{Z}'_{ik1}$ and $\overline{Z}'_{ik2}$ are the means of dummy variables Z_{jk1} and Z_{jk2} for the kth JAR variable, respectively; $\alpha_{ik1}\overline{Z}_{ik1}$ and $\alpha_{ik2}\overline{Z}_{ik2}$ are defined as the estimates of the mean drop on the response (i.e., overall liking) caused by the kth JAR variable being Too Little and being Too Much, respectively. When all attributes of the product are JAR, the intercept γ'_0 is equal to the predicted mean score of overall liking. When one or more diagnostic attributes are not JAR, the intercept is larger than the predicted overall liking mean score. The difference between the intercept and the predicted overall liking mean score can be interpreted as the total/overall mean drop resulting from not being JAR or as the maximum potential improvement margin on overall liking if the attributes that are not JAR are modified to become JAR.

Similar to penalty analysis, important dummy variables in a PLS model can initially be selected according to the prespecified percentage of consumers (e.g., 20%) who rate the attribute not to be JAR and then having the jackknife optimization method applied to select the final important dummy variables. To compare with penalty analysis, it is recommended that dummy variables with 20% or more consumers scored be used in the PLS model.

10.6.3. Analysis Example

The same sample data set used in section 10.5 is used here.

10.6.3.1. Correlation Analysis

Correlation analysis is very important for determining which model (covariance analysis or PLS model) is appropriate for determining the relationship between liking and diagnostic attribute ratings. This is because covariance analysis requires the assumption of independence among JAR variables. Correlation coefficients among the five original JAR variables given in the American Society for Testing and Material data set (Size, Color, Flavor, Thin/Thick, and Stickiness) are presented in Table 10.9. The maximum correlation coefficient was -0.4043 between Thin/Thick and Stickiness, which was significantly different from zero, with $p < 0.0001$, and the remaining correlation coefficients were less than ± 0.12. Because some of the five JAR variables were correlated with each other,

Table 10.9. Pearson's correlation coefficients between JAR variables.

	Size	**Color**	**Flavor**	**Thin/Thick**	**Stickiness**
Size	1	0.0004	−0.0040	−0.0039	0.0947*
Color		1	0.0255	−0.1115*	0.0163
Flavor			1	−0.0280	0.0648
Thin/Thick				1	−0.4043**
Stickiness					1

*Significant at $p < 0.05$; **significant at $p < 0.0001$ ($n = 102$).

the independence assumption for the analysis of covariance was somewhat violated. In this case, it is more appropriate to use the PLS or PCA model than to use the covariance analysis model. The following section focuses only on use of PLS models with dummy variables to map relationships between JAR and liking variables for each individual sample.

10.6.3.1. Partial Least Squares Regression with Dummy Variables

Five pairs of dummy variables for the five original JAR variables were created using the scheme for a 5-point JAR scale in Table 10.8.

PLS regression models using the five pairs of dummy variables were separately fitted by the Unscrambler software program to the data for each individual product. The jackknife optimization method was applied to the PLS models and allowed the identification of dummy variables significantly affecting the prediction of overall liking. The fitted PLS models including only significant dummy variables are presented in Table 10.10.

For sample 170, the jackknife method resulted in the selection of a single pair of dummy variables (Z_{51}, Z_{52}) representing stickiness. For the pair (Z_{51}, Z_{52}), the regression coefficient of 1.2692 (which stands for the absolute value hereafter) for Z_{52} means that the overall liking (OL) mean score decreased at a rate of 1.2692 per unit increase in JAR stickiness scores over the region of 3–5. The regression coefficient for Z_{51} was zero, indicating that the OL mean score was not affected by Z_{51}. The combined effects of the pair of dummy variables (Z_{51}, Z_{52}) for stickiness on OL are graphically presented in Figure 10.8. The figure shows that the OL mean score was constant at 5.7436 (regression intercept) over the JAR region of 1–3 for stickiness and dropped linearly over the JAR region of 3–5, implying that "too sticky" was more detrimental to OL than "not sticky enough." The estimated mean drop on OL for this sample was 0.224 [= 1.2692 × 0.1765, where 0.1765 was the mean of Z_{52}; see Equation (10.28)] because of the product being "too sticky." The regression intercept was 5.7436, indicating that the potential maximum mean score was 5.7436 if all JAR attributes were overall JAR. The difference between the intercept and the predicted OL mean was 0.224, indicating that the maximum potential improvement margin on OL was 0.224 if the attributes that were not JAR were modified to be JAR. Although the PLS model did not give a very good prediction of the OL scores for individual consumers (i.e., low R value in Table 10.10), it predicted well the observed OL mean score (Table 10.10) for sample 170. As far as consumers were concerned, 17% of the consumers rated this product "too sticky," and 1% of the consumers rated it "not sticky enough." Similar to penalty analysis, if only attributes for which 20% or more consumers found the attribute intensity not to be JAR were initially selected into the PLS model, no attributes were found by the jackknife method to have significant effects on OL in the final PLS model (results not shown here).

For sample 458, three important pairs of dummy variables, (Z_{31}, Z_{32}) for flavor, (Z_{41}, Z_{42}) for thin/thick, and (Z_{51}, Z_{52}) for stickiness, significantly affected OL (Table 10.10) because at least one of two paired regression coefficients was not zero. For the flavor attribute, regression coefficients for the pair (Z_{31}, Z_{32}) were 0 and 0.6716, respectively, indicating that "too strong" of a flavor had a negative impact on OL. The product was rated as "too strong" for flavor by 32% consumers, whereas only 6% of consumers rated it as "too weak." The estimated mean drop on OL was 0.283 (= 0.6716 × 0.4216, where 0.4216 was the mean of Z_{32}). For the stickiness attribute, only 4% and 14% of consumers rated the sample "not sticky enough" and "too sticky," respectively. If the 20% were to be

Table 10.10. Results from PLS models using dummy variables for five samples.

		170		458		523		896		914	
JAR Attribute		Percentage of Panelists	Estimate	Percentage of Panelists	Estimate	Percentage of Panelists	Estimate	Percentage of Panelists	Estimate	Percentage of Panelists	Estimate
Intercept			5.7436		6.1246		6.4828		6.5282		7.4183
Size	Z_{11}	24	0	32	0	25	0	22	0	27	0
	Z_{12}	40	0	32	0	39	0	39	0	41	0
Color	Z_{21}	11	0	13	0	11	0	1	0	21	0
	Z_{22}	6	0	1	0	5	0	3	0	0	0
Flavor	Z_{31}	16	0	6	0	5	0	5	0	53	0.8151
	Z_{32}	23	0	32	−0.6716	22	−0.8067	13	0	3	0
Thin/Thick	Z_{41}	13	0	16	0.8586	15	0.7670	9	0	4	0
	Z_{42}	5	0	9	−0.8737	3	0	3	0	21	−1.2391
Stickiness	Z_{51}	4	0	7	0	2	0	1	0	10	1.1624
	Z_{52}	17	−1.2692	20	−0.7837	16	−0.9864	14	−1.5249	9	0
R		0.24		0.47		0.46		0.30		0.53	
RMSE		2.13		1.52		1.54		1.82		1.77	
N		102		101		102		102		101	
Observed OL mean		5.52		5.48		6.00		6.30		6.51	
Predicted OL mean		5.52		5.47		6.00		6.30		6.55	
Rank of preference		4		5		3		2		1	
Potential Improvement		0.224		0.655		0.483		0.228		0.868	

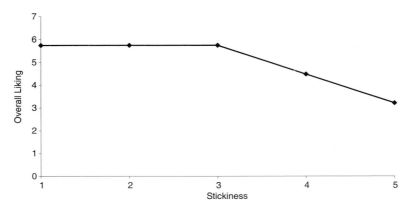

Figure 10.8. Effect of stickiness on overall liking for sample 170.

applied here, it would be concluded that stickiness has no effect on the acceptance of the product. If the rule is not applied, the PLS model shows that only the dummy variable Z_{52} for stickiness significantly decreased the OL mean score, indicating that the "too sticky" texture was detrimental to OL. The estimated mean drop on OL was 0.161 (= 0.7837 × 0.2059, where 0.2059 was the mean of Z_{52}). For the thin/thick attribute, two regression coefficients (0.8586 and 0.8737) for Z_{41} and Z_{42} were significantly different from zero, indicating that there were possibly two segments of consumers: one group would have liked the product to be thicker, whereas another group would have liked the product to be thinner. Because a regression coefficient of 0.8737 for Z_{42} was slightly larger than regression coefficient of 0.8586 for Z_{41}, the paired t-test [Equation (10.27)] was conducted to test whether the two regression coefficients were the same or not. It was found that there was no significant difference in RMSE value between F-model (containing Z_{32}, Z_{41}, Z_{42}, Z_{52}) and R-model (containing Z_{32}, $Z_4 = Z_{41} - Z_{42}$, and Z_{52}) at $\alpha = 0.05$, indicating no significant difference between the two regression coefficients. It was concluded that "too thin" and "too thick" scores statistically decreased OL at the same rate and that the R-model was more appropriate for sample 458 than the F-model. The results from R-model are presented in Figure 10.9a, which is similar to the graphic presentation of the results from penalty analysis. The figure is the plot of the mean drop on OL as a result of its not being JAR versus percentage of consumers who rated the product not to be JAR. For a dummy variable, overall mean drop on overall liking is the product of the regression coefficient by the mean of the dummy variable across consumers. Figure 10.9a shows that "too strong" of a flavor caused the most mean drop on OL, whereas "too thick" of a texture caused the least mean drop for this sample. Although the drop rates for "too thin" and "too thick" were the same, "too thick" texture dropped the OL mean score more than "too thin" texture because more consumers scored the product "too thick" than "too thin." This sample was the least liked of the products included in this data set in terms of the predicted OL mean scores.

For sample 523, the jackknife method selected three pairs of dummy variables: (Z_{31}, Z_{32}) for flavor, (Z_{41}, Z_{42}) for thin/thick, and (Z_{51}, Z_{52}) for stickiness (Table 10.10). The regression coefficients indicated that "too strong" flavor had more negative influence on OL than "too weak" flavor, "too thick" texture was more detrimental to OL than "too thin" texture, and "too sticky" texture was also more detrimental to OL than "not sticky enough" texture. The mean drop plot (Figure 10.9b) shows that more consumers rated the flavor as

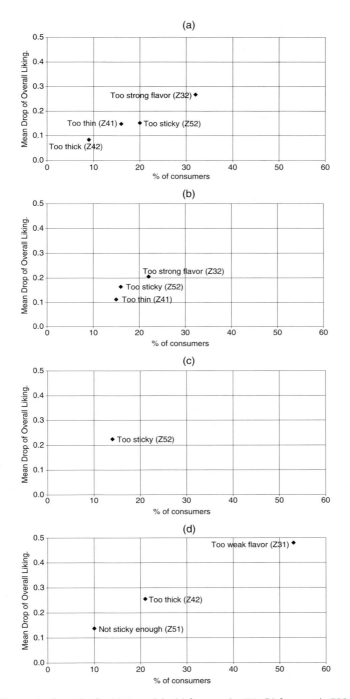

Figure 10.9. The results from the final PLS models: (a) for sample 458, (b) for sample 523, (c) for sample 896, and (d) for sample 914.

"too strong" for this sample and the "too strong" flavor ratings were more responsible for decreasing overall liking scores than "too thick" or "too sticky" ratings.

For sample 896, only one pair of dummy variables (Z_{51}, Z_{52}) was selected by the jackknife method for stickiness (Table 10.10). Dummy variable Z_{51} had no significant effect on OL, but dummy variable Z_{52} significantly decreased OL (Table 10.10). This implies that "too thick" ratings were detrimental to OL, whereas "too thin" ratings had no effect on OL. The mean drop plot (Figure 10.9c) shows that only 14% of consumers rated the product as "too sticky" and that the drop of the OL mean score was 0.224. It is also interesting to see that although a large percentage of consumers found the size of the product either "too small" (i.e., 22%) or "too large (i.e., 39%), the non-JAR ratings did not have a significant effect on overall liking.

For sample 914, the jackknife method determined three pairs of dummy variables—(Z_{31}, Z_{32}) for flavor, (Z_{41}, Z_{42}) for thin/thick, and (Z_{51}, Z_{52}) for stickiness (Table 10.10)—to be important drivers of overall liking. A comparison of the regression coefficients for each pair of dummy variables shows that a "too weak" flavor, a "too thick" texture, and a "not sticky enough" texture were all detrimental to OL. The mean drop plot (Figure 10.9d) shows that 53% of consumers rated the flavor of this sample "too strong," with a corresponding mean drop of 0.479; 21% of consumers scored the texture of the product "too thick," with an estimated mean drop of 0.255; and 10% of consumers scored the texture as "not sticky enough," with an estimated mean drop of 0.138. Figure 10.9 shows a trend toward a mean drop on OL increase as the percentage of consumers rating the attribute as not JAR increased. This is because the means of the dummy variables are dependent on the number of consumers who rate the variables as not JAR. It is evident from Table 10.10 that sample 914 had the highest regression intercept (7.4183) and predicted OL mean score (6.55). This sample was the most liked product, with an observed OL mean score of 6.51.

In summary, the dummy variable method is very flexible and can be used with many regression models. If the JAR variables are independent from each other, dummy variables are used with covariance analysis models for modeling overall liking from diagnostic attributes. If the JAR variables are correlated with each other, dummy variables are used with PCR or PLS models for mapping relationships between JAR and liking variables. The method also provides us with a tool to perform various hypothesis tests. Similar to penalty analysis, it uses a similar graphic presentation of the relationships between JAR and liking variables, but unlike penalty analysis, it is a regression method, which can estimate the "true" mean drop on overall liking caused by an attribute not being JAR. The predicted OL mean scores for products can be used to identify the most liked products. The difference between the regression intercept and actual OL mean score indicates the average potential improvement margin if the product is modified to adjust all diagnostic attributes to a JAR level.

The drawback of this method is that there is no single software program to implement it. SAS can provide most of the calculations, but there is no simple method to implement jackknife optimization that is available for PLS models under the SAS environment (see appendix 10.3 for an example). To implement variable selection methods such as the jackknife method, multivariate software such as Unscrambler will need to be used. A Microsoft Excel macro is available from the Web site associated with this book to automatically create the dummy variable scores. Details about the use of Unscrambler are provided in previous chapters.

We recommend that correlation analysis of JAR variables be performed before modeling the relationship between hedonic and diagnostic scores to determine whether the independency assumption of the JAR variables is met. If the assumption is met, it is recommended that one use the covariance analysis model with dummy variables because the effects of treatments/products, consumer panelists, and other factors on OL can be tested simultaneously. Otherwise, principal components-based regression models (such as PLS or PCR models) with dummy variables are recommended.

References

Agresti, A. 1990. Categorical data analysis. New York: Wiley.
Craven P., Wahba G. 1979. Smoothing noisy data with spline functions: estimating the correct degree of smoothing by the method of generalized cross-validation. Numer Math 31:377–403.
Efron, B. 1982. The jackknife, the bootstrap, and other resampling procedures. Philadelphia: SIAM Publications.
Friedman, J. 1991. Multivariate adaptive regression splines. The Annuals of Statistics, 19:1–141.
Gimenez, A.M., Gambaro, A., Varela, P., Garitta, L., Hough, G. 2003. Use of survival analysis methodology to estimate shelf-life of "alfajor". Proceeding of 2003 Pangborn Sensory Evaluation Meeting, Boston, USA.
MARS. 2001. MARS User's Guide, Salford Systems.
Meullenet, J.-F., Xiong, R., Hankins, J.A., R., Dias, P., Zivanovic, P., Monsoor, M.A., Bellman-Homer, T., Liu, Z., Fromm, H. 2003. Preference modeling of commercial toasted white corn tortilla chips using proportional odds model. Food Quality and Preference: 14 (7): 603–614.
Moskowitz, H.R., Jacobs, B.E., 1987. Consumer evaluation and optimization of food texture. In " Food texture: instrumental and sensory measurement", Dekker, New York.
Sephton P. 2001. Forecasting recessions: can we do better on MARS™? Federal Reserve Bank of St. Louis Review, March/April 2001, 83(2):39–49.
Snedecor, G.W., Cochran, W.G. 1976. Statistical methods. Ames, Iowa: The Iowa State University Press.
Stevens, S.S. 1975. Psychophysics: an introduction to its perceptual, neural, and social aspects, John Wiley, New York.
Xiong, R., Meullenet, J.-F. 2004. Application of Multivariate Adaptive Regression Splines (MARS) to the Preference Mapping of Cheese Sticks. In review, J. Food Sci.

Appendix 10.1 Bootstrapping Penalty Software

- Prepare the data file in an Excel format. The raw data must be used for the program. All the variable names must be in the first row of the data file.
- Install the program through add-ins: Open Excel Program → Tools → Add-ins → Select BootstrappingPenaltyAnalysis (use browse to find it if necessary) → OK. A menu called "Penalty Analysis" will appear in the Excel menu bar.
- Open your data file in Excel Window.
- Run the program. Select "Penalty Analysis" menu item in the "Penalty Analysis" menu and the Penalty Analysis Window will appear. Select variables for Liking Attribute, JAR attributes and Product ID (see Figure 10.5). Proceed by pressing OK.
- Select options for the bootstrapping method and output.
- The program provides results from both penalty analysis and bootstrapping penalty analysis.

Appendix 10.2 Using MARS

To prepare the data file for MARS, raw data must be used. MARS can open various types of files including Excel and text files. However, all the variable names must be in the first row of the data file and should not contain spaces or special characters and should be a maximum of eight characters in length.

Opening the data file in MARS is very similar to opening files in Microsoft Office–based programs. File → Open → Data File → Open File Dialog shows up, and then select the data file → OK. MARS will indicate whether the file was successfully opened.

- Set up the model. Model → Set Up Model → Setup Model Dialog appears, select the response/target variable (Overall liking in this example) and predictive variables (JAR size, JAR color, JAR flavor, JAR thin/thick, and JAR stickiness in this case) for the MARS model.
- Select options if necessary. There are many options available in MARS (see MARS help file for more details).
- Run the analysis by clicking the Best Model button.

MARS provides a summary of the fitted model. For more detailed information, click the Curves and Surfaces button to see graphical relationships between the response variable and predictive variables. Click the Basis Functions button for the basis functions and models.

Appendix 10.3 Fitting the Proportional Odds Model to the American Society for Testing and Materials Sample Data Set

```
Proc logistic data=CaseStudy;
     Class Sample (ref="170")/param=ref;
     Model Flavor=Sample/link=logit scale=none aggregate;
     Title "Proportional odds model for Flavor";
     Contrast   "Samples  458"  vs  "170"  Sample  1  0  0  0
/estimate=both;
     Contrast   "Samples  523"  vs  "170"  Sample  0  1  0  0
/estimate=both;
     Contrast   "Samples  896"  vs  "170"  Sample  0  0  1  0
/estimate=both;
     Contrast   "Samples  914"  vs  "170"  Sample  0  0  0  1
/estimate=both;
     Contrast   "Samples  523"  vs  "458"  Sample  -1  1  0  0
/estimate=both;
     Contrast   "Samples  896"  vs  "458"  Sample  -1  0  1  0
/estimate=both;
     Contrast   "Samples  914"  vs  "458"  Sample  -1  0  0  1
/estimate=both;
     Contrast   "Samples  896"  vs  "523"  Sample  0  -1  1  0
/estimate=both;
     Contrast   "Samples  914"  vs  "523"  Sample  0  -1  0  1
/estimate=both;
     Contrast   "Samples  914"  vs  "896"  Sample  0  0  -1  1
/estimate=both;
Run;quit;
```

Appendix 10.4 Fitting the Proportional Hazards Model to the American Society for Testing and Materials Sample Data Set

```
Proc logistic data=CaseStudy;
    Class Sample (ref="170")/param=ref;
    Model Flavor=Sample/link=cloglog scale=none aggregate;
    Title "Proportional hazards model for Flavor";
    Contrast   "Samples  458"  vs  "170"  Sample  1  0  0  0
/estimate=both;
    Contrast   "Samples  523"  vs  "170"  Sample  0  1  0  0
/estimate=both;
    Contrast   "Samples  896"  vs  "170"  Sample  0  0  1  0
/estimate=both;
    Contrast   "Samples  914"  vs  "170"  Sample  0  0  0  1
/estimate=both;
    Contrast   "Samples  523"  vs  "458"  Sample  -1  1  0  0
/estimate=both;
    Contrast   "Samples  896"  vs  "458"  Sample  -1  0  1  0
/estimate=both;
    Contrast   "Samples  914"  vs  "458"  Sample  -1  0  0  1
/estimate=both;
    Contrast   "Samples  896"  vs  "523"  Sample  0  -1  1  0
/estimate=both;
    Contrast   "Samples  914"  vs  "523"  Sample  0  -1  0  1
/estimate=both;
    Contrast   "Samples  914"  vs  "896"  Sample  0  0  -1  1
/estimate=both;
Run;quit;
```

Index

Adjacent categories model, of ordinal logistic regression, 144
Agglomerative hierarchical cluster analysis, 7
American Society for Testing and Materials data set, 234–235
Analysis of variance. *See* ANOVA
ANOVA (analysis of variance)
    dummy variables and JAR data, 221–224, 221*eqs.*, 222*fig.*, 222*table*, 223–224*eqs.*
    liking level differences measurement by, 6
    PanelCheck panel performance measure, 31
    panelist and panel performance measure, 40, 42–43*figs.*
    proportional odds model (POM), and 148
    t-tests/ANOVA, of significant differences among JAR scores distributions, 215
Assessing the heterogeneity of consumer responses. *See* Consumer testing

Bayesian Information Criterion, 123
Binary logistic regression
    contingency table, 133–135, 134–135*eqs.*, 134*table*
    contingency table, combined, 135, 135*table*
    defined, 133, 133*eq.*
    examples of, 139–144, 139*eq.*, 141*eq.*
    goodness-of-fit test statistics, 137–138, 137–138*eqs.*
    grouped *vs.* ungrouped binary/categorical data, 133–135, 134–135*eqs.*
    individual liking/disliking results, 133, 134*table*
    linear logit regression model, 136, 136*eqs.*
    logit model estimation, 136–137
    model diagnostics, 138
    multicollinearity, 138–139
    overdispersion, 138
    probability of success, 135
Bliss point concept, of JAR data, 207
Bootstrapping penalty analysis, 5*table*
    attributes which penalize acceptance, 210–211, 211*eqs.*
    liking attributes measure, 7
    replications focus, 210–212, 211*fig.*
    software instructions, 233

Chi-square McNemar test, of significant differences among JAR scores, 215
Chocolate chip cookies data set, 25, 40, 40*fig.*, 112–113
Cluster analysis consumer segmentation technique
    Euclidian distance measure, 111–112
    similarity or distance matrix, 111–112
    variable standardization and, 112
    Ward's criteria method, 112, 112*eqs.*, 114, 121*fig.*, 122*fig.*
    See also Hierarchical cluster analysis, of consumer segmentation
Clustering-of-variables approach, to consumer segmentation, 125–126, 125*eqs.*, 126*eqs.*
Consumer segmentation techniques
    cluster analysis, 111–112, 112*eqs.*
    clustering-of-variables approach, 125–126, 125*eqs.*, 126*eqs.*
    fuzzy clustering method, 125
    hierarchical cluster analysis, 113–126

Consumer segmentation techniques (*cont.*)
  identify sensory attributes driving liking focus, 111
  introduction regarding, 111
  LC models, 112–113
  muscadine grape juices data set, 18–19, 21
  See also Hierarchical cluster analysis, of consumer segmentation
Consumer testing
  agglomerative hierarchical cluster analysis, 7
  of fried mozzarella cheese stick appetizers data set, 21, 23, 23*table*, 24*table*, 213, 214*fig.*, 214*table*, 215
  hedonic scores, distances to ideal points, 7
  heterogeneity assessment of, 6–7
  internal preference mapping, 5*table*, 6
  Landscape Segment Analysis, 6–7
  liking level differences, 6
  mapping of consumer ideals concept, 6–7
  MDPREF measure, 4*table*, 6
  move towards, 6
  on muscadine grape juices data set, 19, 21
  PREFMAP framework, 5*table*, 6
  product sensory characteristics and, 6
  response surface methodology, 7
  sensory attributes driving product liking, 7
  of white corn tortilla chips, 15, 16*table*, 17–18*table*
  See also Consumer segmentation techniques; Liking level differences, in consumer testing
Continuation ratio model, of ordinal logistic regression, 144

Data sets
  for panelist and panel performance evaluation, 25
  on web site, 8
  See also Fried mozzarella cheese stick appetizers data set; Muscadine grape data set; White corn tortilla chips data set

Deterministic extensions to preference mapping
  conclusions regarding, 89–90, 93–94
  deterministic model-based stochastic model method of QRA, 164–165
  introduction regarding, 69–70
  See also Euclidian distance approach measure; PLSR (partial least squares regression) on average data; Response surface model (RSM) for external mapping
Deviance and Pearson Goodness-of-Fit Statistics Table, 151
Diagnostic scale data. *See* JAR (just about right) data
Differences in liking level. *See* Liking level differences
Distributions of sensory scores, in liking attributes measurement, 7
Dummy variables and JAR data
  analysis of covariance with, 221–224, 221*eqs.*, 222*fig.*, 222*table*, 223–224*eqs.*, 232
  correlation analysis with, 227–228, 227*table*, 232–233
  limitations of, 232–233
  mean drop on overall liking concept, 220–224, 222*fig.*, 222*table*, 228, 229*table*, 230, 231*fig.*, 232
  PLSR with, 224–227, 224–227*eqs.*, 228, 229*table*, 230, 230*fig.*, 231*fig.*, 232–233

EDCM. *See* Euclidian distance approach measure
Eggshell Plot panel performance measure, 31
EPM. *See* Preference mapping, external
Euclidian distance approach measure, 5*table*
  in cluster analysis, 111–112
  consumer external preference map in sensory space, 87–89, 88*fig.*
  contour map of consumer liking, 89, 92*fig.*
  ideal point correlation, 89
  RSM method comparison, 89, 91*fig.*, 93–94, 93*table*

vs. LSA optimal sensory profile, 103–104, 104*table*, 105–106, 105–107*figs.*
  See also Hierarchical cluster analysis, of consumer segmentation
European Sensory Network multicountry projects, panel assessment, 44–45
European Union ProfiSens project, 45
External mapping
  limitations of, 89–90, 93–94
  See also Euclidian distance approach measure; PLSR (partial least squares regression) on averaged data; Preference mapping, external (EPM); Quantitative risk assessment (QRA); Response surface model (RSM) for external mapping
External preference mapping, of consumer responses, 7

Feedback calibration, 32, 46
Free Choice Profiling, 37
Fried mozzarella cheese stick appetizers data set
  consumer hedonic, diagnostic, and purchase intent means for, 24*table*
  consumer testing, 21, 23, 23*table*, 24*table*
  descriptive analysis, 23, 25
  hedonic and diagnostic means for, 22*table*
  introduction regarding, 21
  MARS analysis of, 213, 214*fig.*, 214*table*, 215
  samples, 21
  scales used in consumer testing of, 23*table*
  summary results, 24*table*, 25
Fuzzy clustering method, of consumer segmentation, 125

Generalized procrustes analysis. *See* GPA
Goodness-of-fit test statistics logistic regression model, 137–138, 137–138*eqs.*, 155, 157
  in MARS, 192
  POMs *vs.* proportional hazard models, 217*table*
  in quantitative risk analysis, 166

GPA (generalized Procrustes Analysis), 4*table*
  asparagus data set, 40, 42*fig.*
  chocolate chip data set, 40, 40*fig.*
  dimensionality concept, 32
  European Sensory Network multicountry projects, 44–45
  Free Choice Profiling, 37
  INRA approach, 45–46
  lemon data set, 40, 43*fig.*
  for lexicon reduction, 43–44
  panel homogeneity measure, 3
  panel-to-panel consonance technique, 6
  panelist, panel performance measure, 37–44*figs.*, 40, 43, 46
  scaling effect, 31
  wine evaluation data set, 37–39*figs.*, 40, 41*fig.*, 42*fig.*, 43, 44*fig.*

Hierarchical cluster analysis, of consumer segmentation
  centered preference data used, 114, 116*table*, 118–119, 118*fig.*, 120*table*, 122*fig.*
  Euclidian distances affected by data manipulations, 114, 118, 118*fig.*
  Euclidian distances used in, 56, 57*fig.*, 119, 120*table*, 121, 121*fig.*
  external data focused method, 124–125
  landscape segment analysis used in, 119, 121
  phased approach to, 119
  raw preference data used, 113, 114, 115*table*, 118*fig.*, 119, 120*table*, 122*fig.*
  segmentation scheme relevance determination, 113
  standardized preference data used, 114, 117*table*, 118–119, 118*fig.*, 120*table*, 121, 122*fig.*
  vs. latent class models, 4*table*, 121, 122–123, 124*fig.*
  Ward's criteria method used in, 112, 112*eqs.*, 114, 121*fig.*, 122*fig.*

Ideal point concept, mapping of consumer ideals, 6–7
IFPrograms, probabilistic unfolding software, 98, 100, 101, 103

Institute for Perception, Richmond, VA, 98, 100
Internal preference mapping, of consumer responses, 6
See also Preference mapping, internal

Jackknifing resampling method in PLSR, 71, 73, 166, 202
  with dummy variables, 225, 228, 229*table*, 230, 232
JAR (just about right) data, 207
  appropriateness of specific sensory levels, 207
  bliss point concept, 207
  bootstrapping penalty analysis, 5*table*, 210–212
  diagnostic attribute scales, 207, 208*table*
  dummy variables and, 220–221
  dummy variables and, analysis of covariance with, 221–224
  dummy variables and, correlation analysis example, 227–228
  dummy variables and, PLSR with, 224–227, 228–233
  introduction regarding, 207–208
  liking attributes and, 7
  MARS analysis and, 194, 196, 201, 202–205, 202*table*, 203*table*, 204*figs*., 212–215, 213–214*eqs*., 214*fig*., 214*table*
  nonlinear relationship over interval ranges, 212, 215
  penalty analysis, 5*table*, 208–210
  proportional odds/hazards approach, 215–220
  raw data distribution importance, 207, 208*figs*.
  See also Bootstrapping penalty analysis; Dummy variables and JAR data; Penalty analysis; Proportional odds/hazards model
Just about right. *See* JAR (just about right) data

Latent class (LC) models, of consumer segmentation, 4*table*, 5*table*
  common variation source identification, 113
  finite mixture models, 112
  hierarchical cluster analysis and, 4*table*, 121, 122–123, 124*fig*.
  Latent Gold modeling software, 112–113, 123
  LC cluster analysis, 112–113
  LC factor analysis, 113
Latent Gold latent class modeling software, 112–113, 123
LC models, of consumer segmentation. *See* Latent class (LC) models, of consumer segmentation
Lexicon reduction, using GPA, 43–44
Liking level differences, in consumer testing
  analysis of variance measurement of, 6
  diagnostic questions, just about right scales, 7
  external preference mapping, 7
  JAR levels and, 7, 194, 196, 201, 202*table*
  MARS technique, overall liking and diagnostic attributes, 7
  modeling from distributions of sensory scores, 7
  modeling from partial least squares regression, 7
  optimal products and, 8
  partial least squares model using average data, 7
  penalty analysis, 7
  projection of external information, 7
  proportional odds model from external data, 7
  sensory intensity factor and, 8
Logistic regression models. *See* Ordinal logistic regression models (OLSR)
Logit, odds, and odds ratio, 130–131*eqs*., 130–133, 131*table*, 132*table*
LSA (landscape segment analysis), 4*table*, 5*table*
  of consumer ideal point mapping, 6–7
  in hierarchical cluster analysis, 119
  optimal product sensory profile, 8
  risk analysis, fried mozzarella cheese sticks, 24*table*, 25
  vs. external mapping methods, 100–101, 102*fig*., 103–104, 103*table*

MANOVA (multivariate analysis of variance), 4*table*, 46
  panelist and panel performance measure, 32–33
  panelists' discrimination among products, panel homogeneity, 3
MARS (multivariate adaptive regression splines), 5*table*
  appearance acceptance, MARS application, 195–196*table*, 196–198, 197*fig.*
  appearance acceptance, PLS regression, 202–203, 203*fig.*
  basis functions concept, 184–185, 184*eqs.*, 184*fig.*, 187, 188, 189–190*table*, 190–191, 191*fig.*, 194, 195–196*table*, 199, 201, 213, 214*table*
  flavor acceptance, MARS application, 194, 194*fig.*, 195–196*table*
  flavor acceptance, PLS regression, 203, 204*fig.*
  fried cheese sticks data set, 25
  generalized cross-validation (GCV), 186–187, 186*eq.*, 188, 189–190*table*, 192, 195–196*table*, 213, 213*eq.*, 214*table*, 215
  goodness-of-fit, 192, 192*fig.*
  high-dimensional data use of, 179
  interaction models, 189–190*table*, 192, 193*fig.*, 194, 195–196*table*, 212–213, 214*table*, 215
  introduction regarding, 179
  JAR (just about right) levels, 194, 196, 201, 202–203, 202*table*, 203*table*, 204–205, 204*figs.*, 212–215, 213–214*eqs.*, 214*fig.*, 214*table*
  knots concept, 182–183, 183*fig.*, 187, 188, 189–190*table*, 190–191, 191*fig.*, 199, 201, 212–213, 214*table*
  liking attributes measure and, 7
  linear combination of basis functions, 179, 191, 194*fig.*, 196
  model quality assessment, 186–187
  overall acceptance, MARS application, 188, 189–190*table*, 190–192, 191*fig.*, 192*fig.*, 193*fig.*, 194
  overall acceptance, PLS regression, 201–202, 201*fig.*
  overall liking and purchase intent relationship, MARS, 181*fig.*, 182, 195–196*table*, 200*fig.*
  overall liking and purchase intent relationship, quadratric polynomial regression, 181–182, 181*fig.*
  penalty concept, 187, 188
  piecewise linear regression splines, 183*fig.*, 185, 191, 215
  PLSR *vs.*, 201–205, 201*fig.*, 202*table*, 203*fig.*, 204*figs.*
  predictor and response variables relationship, 179, 212–213
  purchase intent, MARS application, 195–196*table*, 199, 200*fig.*, 201
  search intensity function, 188
  setting control parameters and refining models, 187–188, 188
  texture acceptance, MARS application, 195–196*table*, 198, 199*fig.*
  texture acceptance, PLS regression, 203–205, 204*fig.*
  two-phase process of, 179, 213
  user instructions, 233–234
Matforsk, Norwegian Food Research Institute, 31
MATLAB, 31
Maximum likelihood (ML) method, of logit model estimation, 136–137
Stuart Maxwell test, of significant differences among JAR scores distributions, 215
MDPREF measure
  of consumer testing, 4*table*, 6
  hierarchical clustering and, 4*table*, 7
  internal preference mapping and, 49, 50, 50*fig.*
  vs. multidimensional scaling, 96, 98
  See also Preference mapping, internal
MDS. *See* Multidimensional scaling
MDSX internal preference mapping software, 50
ML method. *See* Maximum likelihood (ML) method, of logit model estimation

Modeling liking score probability, 7
Monte Carlo quantitative risk
    analysis simulation, 168, 173, 176,
    210
Multicollinearity, in binary logistic
    regression, 138–139
Multidimensional scaling (MDS)
    attribute projection, 100
    badness of fit function concept, 96,
        96eq.
    consumer individuality lost in, 96–97
    consumers not fitting well exclusion,
        101
    derived measures of similarities and,
        95
    drivers of liking identification, 98–100,
        99fig.
    introduction regarding, 95
    landscape segment map creation, 99
    limitations of, 97–98
    low-dimensional space focus, 96
    LSA contour plot, 101, 102fig.
    LSA optimal sensory product
        profile, 103–104, 104table, 105–109,
        107figs., 108fig., 109fig.
    LSA sensory attribute projection,
        observed and projected scores, 101,
        103table
    LSA vs. EDCM, 105–109,
        105–109figs.
    LSA vs. external mapping methods,
        99–101, 102fig., 103–104,
        103–104tables
    LSA vs. internal mapping methods,
        104–109, 105–109figs.
    metric MDS and, 96, 96eq., 97fig.,
        97table
    "nondiscriminators" exclusion concept,
        100–101
    pairwise sample comparisons, 95
    probabilistic unfolding methods,
        98–100, 99fig.
    product sensory properties variation, 95
    similarity models analysis of
        increasing sensory variance, 98–99,
        99fig.
    unfolding and, 96–98, 96eqs., 97table
    vs. external mapping, 95
    vs. MDPREF, 96, 98

Multivariate adaptive regression splines
    See MARS
Multivariate analysis
    MARS technique, 7
    multivariate techniques, 4table
    panelist performance, 4table
    quantification of product differences, 3
    in sensory evaluation, 3
    See also ANOVA (analysis of
        variance); Binary logistic regression;
        Cluster analysis consumer
        segmentation technique; Consumer
        segmentation techniques; Consumer
        testing; Deterministic extensions
        to preference mapping; Euclidian
        distance approach measure; GPA
        (generalized procrustes analysis);
        Hierarchical cluster analysis, of
        consumer segmentation; JAR
        (just about right) data; Liking
        level differences, in consumer
        testing; LSA (landscape segment
        analysis); MANOVA; MARS
        (multivariate adaptive regression
        splines); MDPREF measure;
        Multidimensional scaling (MDS);
        NRV (normalized regression
        vector); Ordinal logistic regression
        models (OLSR); Panelists, panel
        performance; Partial least squares
        regression model; PCA (principal
        component analysis); PLSR (partial
        least squares regression) on average
        data; Preference mapping, external
        (EPM); Preference mapping, internal;
        Preference mapping, of consumer
        responses; PREFMAP framework,
        of consumer testing; Proportional
        odds model (POM), of ordinal
        logistic regression; Quantitative risk
        assessment (QRA); Response surface
        model (RSM) for external mapping
Muscadine grape juices data set
    consumer testing, 19, 21
    descriptive analysis, 19, 20table
    introduction regarding, 18–19
    samples, 19
    sensory lexicon used on, 20table
    summary results, 21, 22table

Musical analogy, of symphony orchestra to panel performance, 27–28

"Nondiscriminators" concept, 100–101
Normalized regression vector.
  See NRV
Norwegian Food Research Institute, 31
NRV (normalized regression vector)
  panel assessment, wine aroma, 32
  panel-to-panel consonance technique, 6
  panelist and panel performance measure, 32, 33–34, 34*table*

Odds, odds ratio, and logit, 130–131*eqs.*, 130–133, 131*table*, 132*table*
Optimal sensory level, explained, 69
Ordinal logistic regression models (OLSR)
  adjacent categories model and, 144
  category intervals are equally spaced concept, 129
  continuation ratio model, 144
  continuous dependent variables focus, 129–130
  equal interval assumption focus, 145
  introduction regarding, 129
  Just About Right scale, 129
  limitations of, 129–130
  logit model estimation, 136–137
  odds, odds ration, and logit, 130–131*eqs.*, 130–133, 131*table*, 132*table*
  order is meaningful focus, 129
  ordinal categorical variable focus, 129
  ordinary least squares regression, limitations, 129–130
  vs. logistic regression, 130
  See also Binary logistic regression; Proportional odds model (POM), of ordinal logistic regression
Ordinary least squares regression, 69
Overdispersion, in binary logistic regression, 138

Paired preference test, 132, 132*table*
PanelCheck panel performance measure, 31
Panelists, panel performance, 6
  consumer testing and, 6
  data sets for, 25
  multivariate techniques, 4*table*
  panel-to-panel consonance techniques, 6
  product assessment reproduction ability, 3
  product discrimination skills, 3
  univariate, multivariate panel performance, 4*table*
Panelists, panel performance: multivariate experience and techniques
  accuracy measure, 46
  attributes in context, 28–29
  conclusions regarding, 46–47
  dimensionality concept, 32
  discrimination measure, 46
  Eggshell Plot measure, 31
  GPA technique, 37–44*figs.*, 40, 43, 46
  GPA technique, European Sensory Network multicountry projects, 44–45
  GPA technique, for lexicon reduction, 43–44
  GPA technique, INRA approach, 45–46
  language of descriptive sensory analysis, 28
  MANOVA technique, 32–33, 46
  musical analogy, 27–28
  normalized RV technique, 32, 33–34, 34*table*
  PCA technique, 37
  PCA technique, chocolate chips, 34, 35–36*figs.*
  PCA technique, wine, 34, 34*table*, 37, 37*fig.*
  replication measure, 46
  sensory evaluation screening and training, 27
  univariate approaches, feedback calibration, 32
  univariate approaches, PanelCheck, 32
  univariate approaches, scatter plots of panelist orientation, 29, 30*fig.*, 31
  univariate approaches, scatter plots of product orientation, 29, 30*fig.*, 31
  univariate approaches, visualizing raw data, 29, 30*figs.*, 31–32
  See also GPA

Partial least squares regression model
  with dummy variables, 7
  in liking attributes measurement, 7
  limitations of, 7
  modeling liking score probability, 7
  using average data, 7
  See also PLSR (partial least squares regression) on average data
PCA (principal component analysis), 4*table*
  dimensionality concept, 32
  multidimensional product representation, 4*table*
  panel homogeneity measure, 3
  panel-to-panel consonance technique, 6
  panelist and panel performance measure, 34, 35–37*figs.*, 37, 40, 41*fig.*
Penalty analysis, 5*table*
  attributes at less than optimal level focus, 209–210, 209*fig.*
  limitations of, 210
  of liking attributes, 7
  See also Bootstrapping penalty analysis
PHM. *See* Proportional odds/hazards model
PLSR (partial least squares regression) on average data
  average external preference mapping, 164
  biplots of product scores and variable loadings, 74–75, 74*fig.*
  collinearity problems solved by, 224
  constant level of one attribute requirement, 70
  data decomposition, 71
  data format, 70–71, 70*table*
  dummy variables and JAR data, 224–227, 224–227*eqs.*, 228, 229*table*, 230, 230*fig.*, 231*fig.*, 232–233
  introduction regarding, 69
  jackknifing resampling method, 71, 73, 166, 202, 225, 228, 229*table*, 230, 232
  leave-one-out validation concept, 73–74
  limitations of, 163
  MARS *vs.*, 201–205, 201*fig.*, 202*table*, 203*fig.*, 204*figs.*
  nonlinear PLSR and, 75
  normalized data, 71, 71*table*
  optimal factors number determination, 73*fig.*, 74
  optimal sensory level, explained, 69
  overall liking *vs.* attribute liking, 69–70
  PLSR method explained, 69–70
  positive *vs.* negative drivers of overall liking, 72
  pseudo full model, 225–226, 230
  pseudo reduced model, 226, 230
  in quantitative risk assessment, 166, 166*eq.*
  ratio of prediction to deviation measure, 74
  regression principle application, 70*table*
  root mean square error, 226
  tortilla data, residual validation variance, factors number function, 73*fig.*, 74
  tortilla data set, observed *vs.* predicted liking scores, 72–73*figs.*, 72–74
  tortilla data set, WRC for attributes, 71–72, 72*fig.*
  See also Quantitative risk assessment (QRA)
POM. *See* Proportional odds model (POM), of ordinal logistic regression
Preference mapping. *See* MARS (multivariate adaptive regression splines); Preference mapping, external (EPM); Preference mapping, internal; Quantitative risk assessment
Preference mapping, external (EPM), 4*table*
  average external preference mapping, 163
  centered *vs.* standardized data use issue, 59–60, 61–63*figs.*
  consumer data fit in the sensory space, 60, 62–64, 64*table*
  extended internal preference mapping and, 65–66, 66*table*
  external data source, limitations, 58–59
  ideal point concept, 63*fig.*, 65
  improvements to, 63*fig.*, 64–65, 64*table*

limitations of, 163
PCA used in, 58, 59–60, 61–63*figs.*,
62–66
popularity of, 59
product locations in two derived
sensory spaces, 60, 62, 63*fig.*
product's sensory characteristics focus,
49, 58
proportional odds model, 157–160,
158–159*table*, 160*fig.*
See also Euclidian distance approach
measure; MARS (multivariate
adaptive regression splines); PLSR
(partial least squares regression)
on average data; Preference
mapping, external; Quantitative risk
assessment; Response surface model
(RSM) for external mapping
Preference mapping, internal, 5*table*,
56–57
conclusions regarding, 66
consumer fit details, 52–53*table*
consumer representation, overall liking
data, 51, 51*fig.*
extended internal preference mapping
and, 65–66, 66*table*
limitations of, 163
LSA *vs.*, 104–109, 105–109*figs.*
MDPREF algorithm, 4*table*, 5*table*,
49–50, 50*fig.*
MDPREF algorithm, consumer
segmentation overall liking, 54,
54*fig.*, 56
MDPREF algorithm, consumer
segmentation preference vector
location, 54, 55*fig.*
MDPREF algorithm, external data
projection into preference space,
54–57, 55*fig.*, 58*fig.*, 59*fig.*
MDPREF algorithm, latent class vector
model, 56–57
MDPREF algorithm, preference-
clustering techniques, 56, 57*fig.*
multidimensional product and consumer
representation, 49–51, 50*fig.*
product component scores, 51,
52*table*
proportional odds model, 154–157,
156*fig.*, 156*table*

singular value decomposition of
centered liking data, 51, 51*table*
See also Deterministic extensions
to preference mapping; MARS
(multivariate adaptive regression
splines); Multidimensional scaling;
Preference mapping, external;
Quantitative risk assessment
Preference mapping, of consumer
responses
consumer segment identification, 49
explanation of, 49
external, 7
individuality response focus of, 49
internal, 6
of muscadine grape juice data set,
18–19
See also Euclidian distance approach
measure; Multidimensional
scaling; PLSR (partial least squares
regression) on average data;
Preference mapping, external;
Preference mapping, internal;
Response surface model (RSM) for
external mapping
PREFMAP framework, of consumer
testing, 4*table*, 6
hierarchical clustering and, 4*table*, 7
See also Preference mapping, external;
Response surface model (RSM) for
external mapping
Principal component regression, 69,
164
Principal component analysis. *See* PCA
Probabilistic unfolding methods, 98–100,
99*fig.*
Probability, odds, and logit, 130–131*eqs.*,
130–133, 131*table*, 132*table*
Product Development Management
Association, 164
Product differences
assessment of, 6
clustering technique, 4*table*
GPA, 4*table*
liking level differences, 6
MANOVA, 4*table*
optimal product attributes and, 8
PCA, 4*table*
proportional odds model measure, 6

Product differences (*cont.*)
  sensory attributes driving product liking, 7
  sensory intensity factor and, 8
  See also Liking level differences, in consumer testing
Product differences, optimal product sensory profile
  barycentric product properties, 8
  LSA projections process, 8
Proportional odds/hazards model
  examples, 217–220, 218*tables*, 219*table*, 230*table*
  POM, parameter estimates, 217, 218*table*
  POM, parameters estimates and odds ratios, 218–219, 218*table*, 220, 220*table*
  POMs *vs.* proportional hazard models, 216, 217*table*
  proportional hazards model, 216, 216*eqs.*, 235
  proportional hazards model, parameter estimates, 219, 219*table*
  SAS codes, 234
  significant differences among JAR scores distributions, 215
Proportional odds model (POM), of ordinal logistic regression, 4*table*, 5*table*, 6, 7
  American Society for Testing and Materials data set, 234
  conclusions regarding, 160
  continuous *vs.* consumer variables, 155
  cumulative density function, 146, 146*eq.*
  cumulative logit model, 146–147, 147*eqs.*, 148–149, 149*eqs.*
  cumulative probability distribution, 146, 146*eq.*
  equal interval assumption focus, 145
  example for, 148
  external preference map, 5*table*, 158–159*table*, 160*fig.*, 164
  goodness-of-fit of model, 151, 155, 157
  internal preference map, 154–157, 156*fig.*
  interpretation ease of, 145
  latent variables and, 145–146, 145–146*eqs.*
  probability density function, 146, 146*eqs.*, 146*fig.*
  SAS statements, 150–151
  testing demographic information effects, 152–154
  testing product differences, 148–152, 152*table*
  See also Proportional odds/hazards model
PROSCAL probabilistic unfolding software, 98

QRA. *See* Quantitative risk assessment
Quantitative risk assessment (QRA)
  average preference mapping through PLS regression, 163
  case study of cheese sticks appetizers, 166–176
  conclusions regarding, 176
  dependent *vs.* independent variable identification, 165
  deterministic model-based stochastic model method, 164–165
  deterministic model development, 164
  introduction regarding, 163
  magnitude of effect of adverse event, 164
  making a decision, 165
  Monte Carlo quantitative risk analysis simulation, 168, 173, 176
  prediction of overall liking, from descriptive sensory attributes, 166, 166*eq.*, 174–176, 175*eq.*, 176*fig.*
  prediction of overall rejection rate, from attribute acceptance data, 166, 167*table*, 168, 168*eq.*, 169*table*, 170–171, 170*table*, 171*fig.*, 172*figs.*, 172*table*, 173
  prediction of overall rejection rate, from descriptive sensory attributes, 173–174, 173*eq.*, 173*table*, 174*fig.*
  preference mapping limitations, 163
  probability of adverse event, 164
  product acceptance score, 164

sensitivity analysis, 171, 172*fig.*, 173
stochastic model development, 165
uncertainty of input variable identification, 164–165
uncorrelated *vs.* correlated input variables, 171, 172*fig.*, 172*table*

Ratio of prediction to deviation measure, in PLSR, 74
Response surface model (RSM) for external mapping, 4*table*, 83–85*eqs.*
  attribute loadings, 75, 76*fig.*
  barycenter method, of optimal product calculation, 82–83, 82*eq.*, 85–86, 86*table*
  compilation of information, 79, 81*fig.*, 82
  generalized inverse matrix, optimal sensory profile calculation, 83–85*eqs.*, 83–87, 86–87*figs.*
  PCA map of sensory space, 75*eq.*
  product scores, 74*fig.*, 75–76
  sensory attribute loadings, 76, 76*fig.*
  sensory profile of optimal product, 82–87, 82*fig.*, 83–85*eqs.*, 86–87*figs.*
  sensory space consumer responses, 76, 78–79, 79–81*figs.*, 82
  Singular Value Decomposition method, 85, 85*eq.*
  tortilla chip data set, sensory space and overall liking, 76, 77–78*fig.*, 78
Risk assessment
  conclusions regarding, 176
  of fried mozzarella cheese sticks data set, 25
  introduction regarding, 163
  See also Quantitative risk assessment (QRA)
@Risk quantitative risk analysis software, 165, 166, 169*table*, 175
RMSE (root mean square error) statistic, 226
Root mean square error (RMSE) statistic, 226
RSM. *See* Response surface model (RSM) for external mapping

SAS internal preference mapping software, 50, 132–133, 139–144, 150–151, 152–154, 155
  proportional odds/hazards model, 216, 217, 234, 235
Schlich, Pascal, 32, 33, 45, 46
SensGear internal preference mapping software, 50, 64
Sensory evaluation
  multivariate analysis in, 3
  See also Panelists, panel performance: multivariate experience and techniques
Senstools internal preference mapping software, 50
Systat internal preference mapping software, 50

t-tests/ANOVA, of significant differences among JAR scores distributions, 215
Temporal Dominance of Sensations method, 28
Tortilla chips. *See* White corn tortilla chips data set

Unfolding MDS models. *See* Multidimensional scaling (MDS)
Univariate analysis
  panelist performance, 4*table*
Unscrambler multivariate statistical software, 71, 74, 166, 202, 225, 228

Ward's criteria method, of hierarchical cluster analysis, 112, 112*eqs.*, 114, 121*fig.*, 122*fig.*
Weighted least squares method
  logit model estimation, 136–137
White corn tortilla chips data set
  appearance measurements, 10, 15, 16*table*, 17–18*table*, 18
  appearance vocabulary for, 14*table*
  consumer testing, 15, 16*table*, 17–18*table*
  descriptive analysis, 10, 11–14*tables*, 14
  flavor measurements, 16*table*, 17–18*table*, 18
  flavor vocabulary for, 10, 13–14*table*, 14

White corn tortilla chips data set (*cont.*)
   introduction regarding, 9
   samples, 9, 10*table*
   sensory properties research on, 9
   summary results, 15, 17–18*table*, 18
   texture measurement, 16*table*, 17–18*table*, 18
   texture vocabulary for, 10, 11–12*table*, 14

White wine descriptive panel data set, 25